Keenie Meenie

Keenie Meenie

The British Mercenaries Who Got Away with War Crimes

Phil Miller

First published 2020 by Pluto Press
345 Archway Road, London N6 5AA

www.plutobooks.com

British Library Cataloguing in Publication Data
A catalogue record for this book is available from the British Library

ISBN   978 0 7453 4078 4   Hardback
ISBN   978 0 7453 4079 1   Paperback
ISBN   978 1 7868 0583 6   PDF eBook
ISBN   978 1 7868 0585 0   Kindle eBook
ISBN   978 1 7868 0584 3   EPUB eBook

Typeset by Stanford DTP Services, Northampton, England
Printed in the United Kingdom

In memory of Vairamuttu Varadakumar,
1949–2019

Contents

Acronyms and Abbreviations

BHC	British High Commission
CDN	Nicaraguan Democratic Coordinator
CIA	Central Intelligence Agency
CJ	Northern Ireland Office files
EPRLF	Eelam People's Revolutionary Liberation Front
EROS	Eelam Revolutionary Organisation of Students
FCO	Foreign and Commonwealth Office
FDN	Nicaraguan Democratic Force, also known as the Contras
IPKF	Indian Peace Keeping Force
IRA	Irish Republican Army
JVP	People's Liberation Front
KMS	Keenie Meenie Services (spellings vary, e.g. Kini Mini, Keeny Meeny, etc.)
LTTE	Liberation Tigers of Tamil Eelam, also known as the Tamil Tigers
MI5	Military Intelligence, Section 5 – Britain's domestic and colonial intelligence agency
MI6	Military Intelligence, Section 6 – Britain's foreign intelligence agency
MOD	Ministry of Defence
PLOTE	People's Liberation Organisation of Tamil Eelam
PREM	UK Prime Ministerial files
RAF	Royal Air Force
RUC	Royal Ulster Constabulary
SAD	South Asia Department (within the FCO)
SAS	Special Air Service
SLAF	Sri Lankan Air Force

Acronyms and Abbreviations

STF	Special Task Force
TULF	Tamil United Liberation Front
WO	War Office files

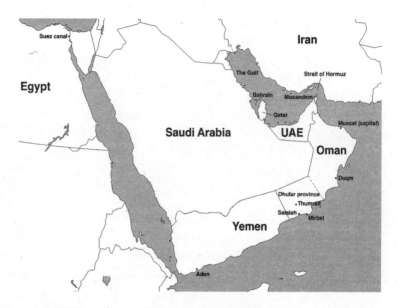

The Arabian Peninsula

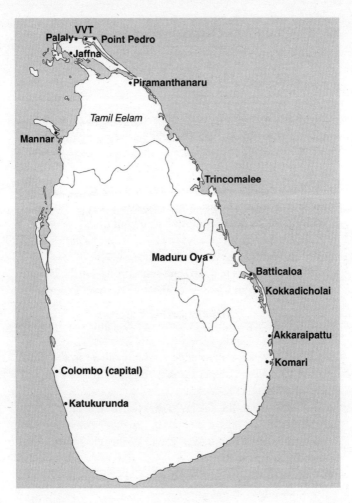

Sri Lanka

Timeline

1975: KMS is founded by Brigadier Mike Wingate Gray, Colonel Jim Johnson, Major David Walker and Major Andrew Nightingale. The company starts guarding British diplomats in Buenos Aires.

1976: Sultan Qaboos of Oman hires KMS to set up and train his special forces. British ambassador to Lebanon hires KMS bodyguards, as does Saudi Arabia's oil minister.

1977: British government fails to pass ban on mercenaries despite Lord Diplock's report.

1978: KMS directors help arrange sponsorship for the National Army Museum in Chelsea. Company's contract in Argentina ends.

1979: UK Foreign Office awards KMS contracts to guard diplomats in Uganda, El Salvador and Rhodesia.

1980: David Walker still listed as a reserve officer in the British army. Thatcher's Cabinet resolves to reduce reliance on KMS bodyguards.

1981: Andrew Nightingale dies in car crash in Oman

1982: KMS guard British diplomats in Uruguay against Argentine threat during Falklands War. David Walker elected as a Conservative councillor.

1983: Sri Lankan government awards KMS contract as country descends into civil war.

1984: KMS start training Sri Lankan police commandos in January. In September, the new unit kills up to 18 civilians at Point Pedro. David Walker starts work in Nicaragua with Oliver North.

1985: Company begins flying helicopters, training army commando unit and commanding operations in Sri Lanka.

KMS bombs hospital in Nicaragua and its personnel in Sri Lanka are linked to torture and disappearances.

1986: KMS attempts to train Afghan Mujahideen in demolition techniques. David Walker steps down as a Conservative councillor. An SAS veteran quits the company over concerns about war crimes in Sri Lanka.

1987: Oliver North testifies before the US Congress about David Walker's work in Nicaragua. Sri Lankan police commandos trained by KMS involved in a massacre of 85 civilians at a prawn farm. KMS pilots give air support to Indian troops amid more massacres of Tamil civilians.

1988: KMS training of Sri Lankan forces scaled back. Sister company Saladin Security becomes increasingly prominent.

Photographs

11. An unidentified KMS instructor with STF recruits; note their M-16 weapons
12. Another unidentified KMS instructor with Special Task Force recruits
13. The STF chief instructors' board records two English names as the first occupants of this important role
14. Athula Dualagala was originally trained by KMS in 1985 and went on to become STF director of training by 2018
15. Joseph Rajaratnam was a maths teacher at Hartley College in Point Pedro when the STF burned down its library in 1984
16. Alasdair MacDermott was a diplomat at the British High Commission in Colombo who recorded allegations of atrocities by Sri Lankan forces linked to KMS
17. Jesuit priest Father John Joseph Mary described atrocities by the STF in Batticaloa, eastern Sri Lanka
18. Lieutenant Colonel Richard Holworthy was defence attaché at the British High Commission in Sri Lanka from 1985 to 1987
19. The author with KMS veteran Robin Horsfall on the roof of his flat in Prague
20. Anthony Knowles after his arrest
21. KMS helicopter pilot Tim Smith
22. A Sri Lankan Air Force Bell 212 carries a national flag at an Independence Day parade in 2019 while a door gunner keeps watch
23. A memorial lists the names of all those who died at the Kokkadicholai prawn farm massacre of 1987
24. Former British high commissioner to Sri Lanka David Gladstone
25. The KMS/Saladin office on Abingdon Road in London
26. Saladin letter to STF Commandant Latiff
27. STF in riot gear advance on protesters, June 2017
28. Grieving relatives hold a portrait of Sathasivam Madisam who drowned while running away from the STF in 2017

Acknowledgements

Special thanks to Khalfan al-Badwawi for his expert research assistance on Oman, and to Lou Macnamara and Rachel Seoighe, particularly for their field work in Sri Lanka. To Angus Frost, not least for coming to Hereford. To Robin Horsfall, Richard Holworthy and David Gladstone for speaking to me. To my friends and family for accepting the long absences I have spent working on this book. I hope now it all seems worthwhile. I am also indebted to Clare Sambrook and Corporate Watch for giving me my first opportunities in journalism and nurturing my writing and research. To Mark Curtis, Ian Cobain, Anne Cadwallader and Abed Takriti for showing me what was possible to unearth at Kew and inspiring me to adopt that method for researching this book. To Daniel Trilling, Lara Pawson and Rebecca Omonira-Oyekanmi for showing me how powerful writing can be. To Matt Kennard, for your wonderful ability to tell stories through towns. To Taimour Lay for your rare blend of investigative journalism and legal knowledge. To Darragh Mackin and Gavin Booth for their guidance on the Peter Cleary case. To Tom Griffin, Kevin Hearty and Rosa Gilbert for their meticulous research on Ireland. To Sam Raphael for sharing our passion of declassified documents. To Yvo Fitzherbert and Bethan Bowett Jones for always being available to read drafts at such short notice. To Jane and Ateeqa for their indispensible assistance. To Greg Walton for his help both at the beginning and end of this journey. To all at Pluto Press for recognising the need for this book to be published. To my Tamil and Sinhalese friends, some of whom it may be safer to mention by first name only. To Ram for planting the seeds and Selven for bringing together my initial research on Ireland, Oman and Sri Lanka. To Virou for his guidance on publishers. To Viraj Mendis, Jude Lal Fernando, Nicolai Jung, Gajendrakumar Ponnambalam, Elil, Terrence, Niraj David, Dr Malathy, Umesh, Ahilan, Agel, Jay, Nalini, Senan, Ravi, Anuraj, Kaj, Mithu, Damian and Santhors. To Karan for his cartography. To so many others who have helped along the way: Antony Loewenstein, Callum Macrae, Andy Higginbottom, Amanda Latimer, Duncan Campbell and Harmit Athwal. To the exiled journalists among you – Bashana Abeywardane and Chandana Bandara in particular, and the fallen ones – such as Sivaram, who I never met but whose writings were the key for understanding British foreign policy. To Sivanandan, for helping me understand better the connections between race, class and empire.

To the archivists at the National Archives, National Army Museum, Surrey History Centre, Shropshire Archives, British Library and of course the SOAS Library. To all the censors who made this process much harder than it should have been, I relished the challenge. To everyone who spoke to me, thank you – and to those who refused, you can still tell me your side of this story.

And ultimately to Vairamuttu Varadakumar, who never lived to read this book, but provided immeasurable support to ensure it would happen.

Prologue

The early morning mist cloaks the village in a false sense of tranquillity. Some hear the helicopters coming. Others slumber too deeply to realise the danger that is hurtling towards them. The air force base is only 20 miles away, so the thump of rotor blades is familiar. Today the noise is louder, a constant crescendo that never stops until the mist is shattered by olive green fuselages probing the villagers' peaceful way of life and altering it permanently.

Scores of soldiers stream out of six helicopters and disappear inside concrete irrigation channels that run like veins through the fields. One helicopter lands next to the house of a young woman, Thurairasa Saradha Devi, and her 21-year-old brother, Ponnuthurai Pakiyanathan. Terrified, Devi runs inside and hides – but soldiers soon surround her home. 'The army ordered us to come out and kneel', she would later recount to a Tamil human rights group.[1] 'There was another child with us who also knelt on the floor.'

Amid the terror, with life and death flashing before her eyes, one memory stood out. Among the helicopter pilots who landed in her village, 'one of them was a tall white man who was watching everything carefully. Many other people in my village saw him that day.' Devi struggled to make sense of this oddity. 'Villagers later referred to him as Mossadu. I didn't know what it meant then. Later I learnt that Mossadu are overseas white men.' Although Israeli security experts were working in Sri Lanka at that time, the pilot she saw was almost certainly not from Mossad. He was a mercenary from a British company, Keenie Meenie

Services or KMS Limited, which had begun flying Sri Lankan Air Force helicopters several months before this incident.

As the white pilot watched over her village, Devi's life began to fall apart. 'Soldiers captured my brother and tied his hands. They took him by the side of the helicopter, made him hold a rifle, and took video footage and a photo. Afterwards they brought my brother to the house and asked me if he was an LTTE man.' The Liberation Tigers of Tamil Eelam was a guerrilla movement fighting to free the island's Tamil minority from the ethnic Sinhalese majority. Devi denied her brother was in the LTTE. 'We are farmers – we are poor people doing farm work here only', she frantically told the soldiers, to no avail. 'We were hit by guns and boots. They said that my brother was LTTE and that they had a photo of him with a gun. If we did not agree with them, they would kill us and all the children. With that they burnt our house down.' More than 75 houses were ablaze. 'There was smoke everywhere', she said. 'We were all shouting and begging for mercy. They took my brother with them. I followed them, crying, and asked the army several times to release him. One soldier kicked me with his boots and I fell on the floor. After some time I opened my eyes. I could not see my brother.'

Other women managed to flee the village before the army surrounded them. One villager ran from lane to lane shouting out warnings. Uma Maheswaran Kamalambihai, 45, took her six children and ran into the nearby forest. Her youngest child was only six months old. They stayed in the forest all day, drinking dirty water from small ponds. But many did not hear or heed the warnings. A survivor later followed the soldiers' boot marks, which stood out from the villagers' modest sandals. The footprints led him to seven corpses.

Devi, the lady who saw the white pilot, fled the village with her children. On their return the next morning, she said: 'We saw so many dead bodies and could not find my brother.' It took them six days to find his body – the smell of her brother's decaying corpse alerted them to its presence. 'There were several stab marks and his hands were tied behind his back. They had stabbed and pushed him from the helicopter. All his bones were broken.' In total, 16 people lay dead, and 30 more were injured. Their wounds included life-changing trauma that left them paralysed or deaf. The soldiers even mugged a farmer, taking his watch and 2,000 rupees before chasing him away. The farmer was so poor that he went back to the army to ask for his possessions. Then the soldiers killed him.

Mothers saw their sons lying dead in pools of blood. Wives watched their husbands beaten by the army – one woman said the bludgeoning was so severe that blood poured out of his ears. Some men were tied upside down to a tree branch and inter- rogated while water was poured down their noses. The village shopkeeper was blindfolded, taken away and executed. Among the almost exclusively Tamil casualties was a 26-year-old Sinhalese civil servant, Vansanatha Kopiyathilaka Kamini. Yet the local army commander would later claim that his troops had only killed Tamil rebels.

The surviving villagers were so poor that they struggled to find coffins for proper funerals. One mother was displaced by the war and cut off from her family. She spent the next two decades not knowing the fate of her son, who was killed in the attack. Among the carnage though, there were some remarkable survival stories. Soldiers tied up 18 villagers and locked them in a room, intending to shoot them later. An eight-year-old boy, who had not been bound, untied all 18 people, allowing them to escape. A local government official, Mr Sinnathamby, also did a valiant job recording the tragedy. He collected details of those killed and injured as well as the properties damaged. The authorities confiscated his notes and interrogated him for three days, and yet he refused to stop speaking out about the massacre. 'These people created Piramanthanaru through their own hard work', he said. 'The army came and destroyed it all.'

Secret British government cables, finally declassified three decades later, support Devi's claim that a white pilot took part in the atrocity. One diplomat said that a mercenary company, KMS, 'appear to be becoming more and more closely involved in the conflict and we believe that it is only a matter of time before they assume some form of combat role however limited it might be'. He noted: 'Members of KMS are frequently seen both at hotel bars in Colombo and at private functions. The identities of many of them (and of the role of the team as a whole) are well known to the British community here.'[2] As 1985 drew to a close, the head of Britain's Foreign Office was informed by his staff that 'We believe only KMS pilots are currently capable of flying armed helicopter assault operations in Sri Lanka.'[3] In time this mercenary air force would deliver decisive blows against the Tamil militants and civilians alike, at immense cost to their liberation struggle and incalculable profit to KMS. The bloodbath at Piramanthanaru that day was not an anomaly, but was set to become part of a gruesome pattern.

Introduction
Return of the Privateers

Profiting from war is one of the most controversial aspects of UK foreign policy. The debate normally centres on why British bombs are being sold to a belligerent ally and how the deal was secured. The recent furore around Saudi Arabia's murder of the journalist Jamal Khashoggi and the ensuing pressure to stop Britain selling more than £4 billion worth of weapons for the war in Yemen is a case in point. However, the arms industry will always defend its business on the grounds that its staff never pull the trigger, and that any subsequent casualties are therefore not its responsibility. Or as the chairman of Britain's largest arms dealer, BAE Systems, modestly told shareholders in 2019: '[We] provide defence equipment that ultimately encourages peace.'[1]

Tenuous as that logic may sound, amid this heated debate on the arms trade it is often forgotten that there is another British industry altogether, which has absolutely no qualms about being directly involved in war. Mercenaries will deliver the bullet directly to their client's target of choice, and the UK has one of the world's largest networks of private armies. There are over a dozen firms clustered near the special forces base in Hereford – corporate warriors who are ready to operate anywhere on earth provided there is profit to be made. As such, British mercenaries can be embroiled in conflicts without much more accountability than their firm filing an annual report at Companies House, where they are often only required to list the region of the planet in which their 'security consultants' operated that year: Africa, Asia or Latin America.

It has not always been this way. The British government did once have a monopoly on violence, as most states aspire to, but this control over bodies of armed men was deliberately relinquished in order to exert power through more subtle and ultimately effective channels. As Britain entered the 1970s, she had lost almost all of her once vast empire, from India to Malaysia, Egypt and Kenya. Former colonies whose riches had paved the streets of London, fed the northern mills with cotton and quenched Britain's thirst with their tea leaves, now governed themselves. As close to home as Belfast, parts of the United Kingdom were in insurrectionary mood and yearning to break away. Even the miners had climbed out of the bowels of the earth with their faces caked in soot and dared to go on strike. For many British people of a certain class, who were born and raised in the colonies, and spent decades in the military or civil service, the decline of empire was deeply disorientating. They had won two world wars, only to spend the next quarter century watching the world decolonise before their eyes.

Of course, they had tried to ensure that the new rulers would be friendly to the old mother country, and play by Britain's rules. It was bearable to give away a country to a puppet ruler who would obediently do their bidding. India was partitioned, her western and eastern flanks carved off, weakening her long-term military potential. The transition to independence in Kenya and Malaya was painfully prolonged, until radical movements like the Mau Mau or Chin Peng's Maoists were crushed. In some cases, Kenyans were literally castrated with pliers.[2] Shocking as they were, these measures were not always enough to stop the former colonies from pursuing complete autonomy from the metropole. In Egypt, the revolutionary pan-Arabist Gamal Abdel Nasser wrenched power from a pliant king, and nationalised the Suez Canal, severing the jugular of the empire's sea lanes. Britannia no longer ruled the waves.

Leaders like Nasser made decolonisation an extremely bumpy ride for the old guard. The stiff upper lip trembled from the earthquake, and constant aftershocks, of ending an empire. By the 1970s, some felt it was time to re-exert power and control before everything was lost. Instinctively, these arch-imperialists reached for a familiar technique, one that had worked so well for their ancestors. They appeared not to have forgotten how the empire began 400 years ago, with Queen Elizabeth I granting permission to enterprising merchants and aristocrats to seek riches around the world. From the privateer Francis Drake and his sorties in West Africa and the Caribbean, to the East India Company's voyages across the Indian Ocean, heavily armed private companies had played a key role in exerting English, and eventually British, power across continents. It was only when this corporate rule was threatened, as in Delhi in 1857 when the East India Company's local troops staged a major revolt, that the full force of the Crown had to step in to crush the uprising. Tens of thousands of European troops who once worked for the company were subsequently absorbed into the Crown's army, and Queen Victoria replaced the firm's directors as the ruler of India. Over a century later, with the world once more in flux and Queen Elizabeth II on the throne, perhaps the reverse was now possible, and necessary? Wherever the Crown retreated, was there now a role for British companies to step in to ensure stability in former colonies? To prevent a communist from taking over? Or a strategic harbour from falling into less reliable hands?

It was in this reactionary climate that Britain's private security industry, as we know it today, began to emerge. Tentatively at first, Special Air Service (SAS) veterans, blooded by Britain's colonial wars, banded together and embroiled themselves in a civil war somewhere foreign. At first they did not even bother to form a company. Men like Colonel Jim Johnson, a former SAS commander, used the basement of his Chelsea home to recruit mercenaries to fight Nasser's forces in Yemen during the 1960s.[3] The setting was so intimate that the participants called it *Beni* Johnson, Arabic for family. That arrangement, which had the sporadic blessing of MI6 and the SAS, was a one-time operation, and dissolved after they had inflicted enough damage on Nasser. It was not until the 1970s that mercenaries such as Johnson would formalise their activities into permanent companies, appoint directors and brand their firms with mysterious names – in Johnson's case, Keenie Meenie Services (KMS) Ltd. Unlike their predecessors, such entities were capable of taking on a series of contracts around the world, and by the 1980s were part of a booming industry, fuelled by free market Thatcherism and relentless privatisation, and supercharged by Ronald Reagan's aggressive anti-communism. These companies were now in their element, sabotaging left-wing regimes as readily as they propped up right-wing dictatorships. Some firms failed to win more than one contract and fizzled out in a few years. Others, those with connections in high places, managed to secure a series of deals that allowed the company to grow and establish itself, so that it was strong enough to weather a storm. If one of its contracts turned sour, the rest of the business could survive, even if it had to rebrand itself occasionally. Their survival meant that, by the time Thatcher left office in the 1990s, Britain had a well-established private security industry. In time, this climate would incubate firms such as G4S whose services, from immigration detention centres and Olympic guards at home, to war zones abroad like Iraq, have become integral to British governments for the last quarter century.

This book is the story of one such pioneering mercenary company, KMS Ltd, whose name has long since faded from the limelight, but whose legacy, and some staff, live on.[4] KMS has left an indelible scar on the 200,000 Tamil refugees scattered across London. I came to know the Tamil community in 2011 while studying politics at the School of Oriental and African Studies in Russell Square. On campus, I was involved in campaigns to stop Tamils being deported to Sri Lanka, where they faced torture. It was amid this drama of placard making and detention centre protests that a Tamil friend told me a childhood story. One day in the 1980s, his uncle said a mysterious British company, KMS Ltd, was helping the Sri Lankan government fight Tamil rebels. This recollection intrigued me. There was very little written about the company – some scattered references to its work in Nicaragua, Oman and Afghanistan, but often more rumour than fact.

After I graduated, I spent several years working as a researcher at Corporate Watch, focusing on the private security industry's role in the detention and deportation of asylum seekers. I spent hours each day mapping out the activities of companies like G4S, Serco and Mitie, but the name KMS stayed in my mind. Slowly, over the last seven years, I have pieced together everything I could find about KMS, especially its work in Sri Lanka but also elsewhere. Primarily, I have relied upon British government files which are, by law, made available to the public at the National Archives in Kew 30 years after they were written. Inside the bowels of this brutalist building in west London there are miles upon miles of once secret documents: telegrams cabled from British diplomats stationed around the world to their headquarters in London, minutes of cabinet meetings held by ministers of the day, and forensic reports written by soldiers analysing distant rebel movements. I combed through British Foreign Office files not only on Sri Lanka, but also from Latin America (Argentina, El Salvador, Nicaragua and Uruguay), the Middle East (Lebanon and Oman), Africa (Angola and Uganda), and Europe (Ireland and the Netherlands). Regimental journals at the National Army Museum in Chelsea proved to be another welcome source, especially for understanding what the mercenaries had previously done when they served in the British military. The Surrey History Centre in Woking, with its meticulous stash of council minutes, shed considerable light on the political life of one KMS founder, who simultaneously worked as a local councillor. US government archives, much of them available online, were a trove of information about the company's work in Nicaragua.

From this formidable array of paperwork, I was able to identify which British officials were involved with KMS, and trace several surviving diplomats who agreed to be interviewed, reuniting

9

them with reports they had written all those years ago. One former defence ministry official was willing to divulge much more detail than the carefully crafted telegrams had revealed, whereas others remained very guarded. Alongside this process, I travelled around Sri Lanka interviewing Tamil priests, lawyers, politicians and widows who had vivid recollections of the company's impact on their struggle for independence. Several colleagues who I worked with also spoke to Sri Lankan military veterans and recorded their fond memories of being trained by British mercenaries. Newspaper clippings at the British library, various film archives and the memoirs of former soldiers have complemented my research considerably.

One aspect of this enquiry was particularly difficult. There has been a considerable delay in the declassification of Foreign Office material about KMS, such is the secrecy and mystique surrounding the firm. The documents I found at Kew were often heavily redacted, with key sentences blacked out or entire pages removed. In dozens of cases the files were simply shredded – or earmarked for destruction until I demanded access to them. This censorship, some of it conducted by the former civil servants who had originally written the telegrams themselves, could only be challenged through freedom of information requests. Essentially, these are emails sent to government departments demanding a right of access to classified material. The law in this area is full of caveats, with anything vaguely related to the special forces or intelligence agencies exempt from disclosure, throwing up a smokescreen around the company's connections to powerful institutions such as the SAS and MI6. And in circumstances where departments do have to hand over documents, they can drag their feet considerably. A delay of six months is not uncommon, even for something as simple as a civil servant posting a document they have already agreed to disclose.

However, when my requests have been successful, thousands of pages of material have arrived in padded brown envelopes. Sometimes, the disclosure was farcical, containing newspaper clippings or parliamentary speeches that have been in the public domain for decades. In other cases, the information Whitehall sought to keep secret was much more sinister, and contained evidence of British complicity in war crimes. Ultimately, the most sensitive material remains classified at the time of writing and is subject to appeals, which may only come to fruition after this book is published. So despite not being able to provide all the pieces of the puzzle, I trust this book contains more than enough to understand the nature of KMS, its complex relationship with the British government and its profound impact on the private security industry.

1

White Sultan of Oman

As the summer of 1970 faded into autumn, a forlorn figure checked in at the Dorchester, a glamorous five-star hotel on London's prestigious Park Lane. The art deco design, marble clad interior and thick carpets were a welcome change from the austere Royal Air Force (RAF) hospital in Wiltshire, where he had spent six weeks convalescing while his bullet wounds healed. Still, putting him in this hotel was just the latest insult that Whitehall could inflict on him. How this tired old man, Said bin Taimur, yearned to be back at his palace in Muscat, where he was the Sultan of Oman. That was until the British, whose interests he had served so faithfully for decades, decided to depose him. They had stormed into his private chamber, disarmed his guards and shot him three times. He tried to fight back, but only managed to shoot himself in the foot. Then a young British army officer, Tim Landon, finally confiscated his pistol, marking the end of his reign.[1]

That man Landon had a lot to answer for. He had shared a room with the Sultan's son, Qaboos, at Sandhurst. The pair got on well at the military academy – Landon protected the Arab prince from senseless bullying. So when Landon turned up in Oman in 1965, the royal family were relieved to see him again. Oman was going through a tough time. There was a revolution brewing in the southern highlands, a border region near Yemen known as Dhufar. This crescent-shaped mountain ridge was quite unlike anywhere else in the arid Gulf country. Dhufar was a rolling plateau, on average almost a thousand metres above sea level. In the monsoon season, clouds laden with moisture cloaked its escarpments, nourishing a belt of dense woodland. The mist was so thick you could hardly see in front of you. The grasslands along the summit made it look more like the Lake District than southern Arabia. The biodiversity in this region was such that botanists filled a book one inch thick with drawings of the different plants, marvelling at their medicinal properties.[2] This unusual, blessed ecosystem made Dhufaris feel different to the rest of Oman. 'Oh Arab people of Dhufar', a communiqué declared in June 1965. 'A revolutionary vanguard from amongst your ranks has emerged, believing in God and in the homeland.' This was the voice of the hitherto unknown Dhufar Liberation Front, proclaiming its existence and ambitions. 'Taking freedom of the homeland as its principle, it has raised the banner of liberation from the rule of the tyrannical Al Bu Said Sultans whose Sultanate has been connected with the columns of British colonialist invasion', the group thundered, echoing the cries of Nasser in Egypt a decade before them.[3]

The people of Dhufar had a proud history of autonomy, and deeply resented being run from Muscat by the Sultan, Said bin Taimur. It was true he had done little for them. Few people in his country could read. Sunglasses were banned, slavery was legal, and there was almost nothing in the way of hospitals or basic infrastructure. There were few schools, and none at all for girls.[4] But at least this ignorance prevented the youth learning about strange foreign ideas, like democracy or communism. Or so the Sultan thought. The trouble was that Dhufar had come under the spell of neighbouring Yemen, where Arab nationalists and Marxists, backed by Nasser's Egypt, had overthrown the old pro-British monarchy. Now the Dhufaris were looking to do the same in Oman. Their movement morphed into the Popular Front for the Liberation of the Occupied Arabian Gulf. It was anti-colonial, anti-caste and anti-class. It stood for gender equality, banning polygamy and female circumcision. The revolutionaries sent girls to school, and women to the battlefield. Slavery was abolished in areas under their control.[5]

Landon was clearly unimpressed by these revolutionary men and women, because when he first showed up in Oman he wanted to keep the Sultan in power. He happily went to work as an intelligence officer in Dhufar. Oman mattered to Britain because of its coastline, especially the northernmost tip at Kumzar, which juts out into the Strait of Hormuz. At its narrowest point, there is just 21 nautical miles separating Oman from Iran – a distance little further than the short ferry ride from Dover to Calais. Through this chicane-shaped channel, a procession of supertankers float a third of the world's seaborne oil out of the Gulf every day, en route to a petrol station near you. In strategic terms, it was and still is another Suez Canal. Landon had found himself in a part of the world that mattered to Britain's imperial interests. Accordingly, he paid particularly close attention to what was happening at the other end of the country, in Dhufar, and schemed about how to contain the unruly highlanders. After a while, he claimed to have identified tribal splits among the opposition, which could be exploited under a divide-and-rule strategy. Collaborators could be incentivised through money, and existing conflicts over land would be manipulated.[6] Apparently the British had done this successfully elsewhere (effectively giving money to traitors to buy back their loyalty), although it seemed a bit elaborate to the ageing Sultan. For years, the RAF had helped him put down rebellions through the aerial bombing of vital lifelines such as wells. The British tactic of 'water denial' in Oman normally put a stop to any unrest.[7] This time it was different though. The revolution in Dhufar was not going away. In fact it was spreading to other parts of the country. Soon it would be in Muscat, at the palace gates.[8] Landon was growing impatient, although the Sultan did not realise it then. Behind his back, the ambitious British officer was plotting a coup with the Sultan's son, Qaboos, tapping into their friendship forged at Sandhurst.

On 12 June 1970, Landon told Whitehall that Qaboos was considering taking power. In a country like Oman, where the British wielded enormous military and political power, deposing one of their most loyal client rulers would have to be carefully stage managed.[9] It took Whitehall weeks of agonising deliberations before the coup was approved. On 23 July, a contingent of English and Irish officers took charge of several Arab soldiers and proceeded to the Sultan's palace in Muscat. Guards were bribed, others surrendered, and it was not long before the plotters had the old Sultan cornered, wounded, disarmed and forced to abdicate. That was how this lonely old man came to check in at the Dorchester Hotel a few months later. Betrayed by his family and foreign 'friends', consigned to live out his final years in obscurity, before they quietly buried him at the Muslim burial ground in Woking. Shortly before his death, he told a confidant that his greatest regret was 'not having had Landon shot'.[10]

* * *

Tim Landon could not have been in a better place at a better time. He had picked the winning side, and backed it from the outset. His foresight was duly rewarded by the new Sultan Qaboos, who made him his equerry, or personal attendant.[11] In this prime position, his counter-insurgency advice would now be listened to and implemented. The so-called 'Landon Plan', to divide and rule the opposition, swung into action. It also included a 'hearts and minds' element. The old Sultan had done little to develop the country. The new one would build basic infrastructure to neutralise his critics' economic grievances. Within months of the coup, defectors were descending from the revolutionary highlands. The British called them Surrendered Enemy Personnel. They were paid, clothed, armed and fed, before being sent back into the mountains under the control of the SAS to fight their former comrades. These deadly defectors were known in Arabic as the *firqat*.[12] Landon's plan was a textbook colonial counter-insurgency strategy. Religion was used to give the defectors a sense of ideological justification for their treachery. The British encouraged them to see their cause as a *jihad* against

15

infidel communists backed by Moscow. Their units were given the names of Muslim conquerors – the first brigade of defectors was called *Salah al-Din*, in other words Saladin. By following Landon's strategy, the new Sultan was able to consolidate his grip on Oman by 1976. The country's Ministry of Information confidently declared that the rebels had 'collapsed', and assured everyone that the armed forces were 'entrusted with the sacred mission of standing firm in the face of terrorist designs backed by hostile powers, and with the sacred duty of defeating their attempts at aggression and sabotage and ridding the country of them and their vile deeds'.[13]

Despite this military triumph, the growing influence of this lone British officer over the new Sultan of Oman irked some in Whitehall. Landon could come across as an establishment insider, a product of empire born on Vancouver Island to an English soldier and a Canadian mother, but he went to school in Eastbourne, not Eton.[14] Now here he was, suddenly a kingmaker. Some even called him the White Sultan. Qaboos was eternally grateful to Landon and showered him with wealth – cheques for £1 million on each birthday according to some;[15] a share of Oman's oil sales others suspected. It was easy to be envious. By the time most of the revolution was crushed in 1976, many British soldiers who had spent years serving the Sultan were being sent home, away from the trappings of a colonial lifestyle, to allow native Omani soldiers to take over. It was called 'Omanization', and some expats were 'unprepared for the extent of change and the ruthlessness of the surgery employed in achieving it'.[16] Landon personally promoted this process – effectively dishing out state jobs to locals to reduce unemployment and the potential for unrest. Of course, he had no intention of going anywhere himself. He was practically an Omani now – Qaboos had given him a passport.

With Landon never far from his side, the new Sultan assid-uously set up a praetorian guard, to prevent his father's fate befalling him. A quarter century later, Western writers would reflect how 'The glaring exception to Omanization of the command structure has been within those forces devoted specifically to the Sultan, such as the Special Forces, Royal Yacht, and Royal Flight Squadrons which are all still under British officers.'[17] Another unit dedicated to the Sultan was the Royal Guard Brigade, whose bodyguards secured his residences. The brigade 'amasses power quickly, apparently untroubled by financial constraints', commented a British Ambassador in Muscat.[18] Should the Guard Brigade fail to protect the Sultan, Royal Flight pilots (many of whom had previously worked for Queen Elizabeth) could presumably drop him on board the Royal Yacht, which 'can act as a safe haven for the ruler and ... possesses a multiplicity of Secure Communications and broadcast facilities.'[19]

It would never reach that point. The Sultan's Special Forces were under his personal command and would nip any unrest in the bud. Even if other branches of the military tried to turn against the Sultan, his Special Forces would stand firm, adorned in their distinctive lavender-coloured berets and belts. They were set up in 1976, modelled on Britain's SAS and trained in their 'special entry techniques'.[20] The unit absorbed many of the *firqat*, now numbering 3,000, when the fighting began to subside.[21] 'Up in the hills of Dhofar [*sic*] the new Special Force is trained by ex-SAS personnel in conditions of extreme extravagance', Britain's ambassador grumbled to the foreign secretary. No expense was spared by the Sultan on this venture. He had hired a company called KMS Ltd, or Keenie Meenie Services, to carry out the job.[22]

No one really knew what this name meant, 'Keenie Meenie'. Perhaps it was Arabic, or Swahili, Maori or even Scottish? One insider claims it was simply a jovial term used by Scots soldiers 'for people who are very keen and nasty at their jobs and very eager to do well'.[23] A Ministry of Defence official who worked closely with the company has told me it was an Arabic expression for 'under the counter', implying that they specialised in clandestine operations – a turn of phrase recognised by one Dhufari

I met who associated it with deviousness and intrigue. General Sir Peter de la Billière, a former director of the SAS, has been quoted as saying 'Kini-Mini, [was] in SAS terms, any undercover activity'.[24] Another possibility comes from Michael Asher, a former SAS solider and accomplished linguist. He suggests that Keenie Meenie is a Swahili phrase describing 'the movement of a snake in the grass', that was borrowed by British officers during the Mau Mau uprising in Kenya to describe the counter-insurgency concept of 'pseudo-gangs'. These were 'white policemen, dressed African-style with faces blackened, [who] accompanied teams of "turned" ex-terrorists into the bush'.[25] The ideas of deception and facade are powerful images for understanding the firm's modus operandi. Was KMS really a private company, or was it just a mask worn by the British establishment to carry out deniable operations?

Whatever the name meant, and many Swahili speakers I consulted are also puzzled by it, Tim Landon was an 'associate' in this mysterious firm, and so it could be trusted with keeping the Sultan in power.[26] Or at least as one civil servant would later put it, Landon 'appears to be associated with KMS even if he may not be a member of the Board of Directors'.[27] The arrangement between Landon and KMS was so secret that the White Sultan would take it to his grave. When he died from lung cancer in 2007 at the age of 64, he was one of Britain's largest landowners, leaving his son Arthur a £200 million inheritance, making him the richest young man in the UK and a friend of Princes William and Harry. Despite amassing so much wealth, his involvement with KMS remained hidden. By then, KMS had rebranded as Saladin Security, perhaps named after the first group of Dhufari defectors. The company provided Landon with bodyguards up until his passing away.[28]

Landon's legacy lives on. His friend since Sandhurst, Sultan Qaboos, still sits on the throne in Muscat. At the time of writing, his reign is headed for a half century, earning him the dubious distinction of being the longest-serving ruler in the Middle East.

Such is the Sultan's fear of unrest, Oman currently spends more money per head on its military than any other country on earth, using its oil wealth to keep a lid on dissent. Although his humble origins may have piqued his classier colleagues, Landon's work in Oman is widely seen as a success story among the British military and diplomatic elite. By the mid-1970s, a fledgling Arab revolution was in reverse, and a reliable Anglophile Sultan installed on the throne. Other former colonies, however, were proving far more troublesome for the old guard.

*   *   *

London, 1976

Tucked away down a sleepy and exclusive west London side street is 11 Courtfield Mews, a small terrace in Kensington. Easily mistaken for a residential flat, it was perfect cover for the brigadier who worked there. Mike Wingate Gray had left the army a few years earlier. Now in his mid-fifties, he still had boundless energy. Founding KMS Ltd with a few old buddies seemed like the perfect way to finish his distinguished career of serving Queen and country. He was battle-hardened, there was no doubting that, but also smart. A veteran of the Normandy landings, he spoke four languages and went on to command the SAS.[29] In the 1960s he had deployed SAS squadrons to Aden and Radfan, parts of Yemen still under British occupation, where they worked in tandem with Colonel Johnson's maverick mercenaries stationed further north in areas already lost to rebels.

The two men had eventually decided to become business partners, and their order book was beginning to pick up as conflicts spread across the Middle East, far beyond Oman and Yemen. On the shores of the Mediterranean, Beirut was rocked by civil war between rival religious and ethnic factions. So when the phone rang at Courtfield Mews, the old brigadier was not surprised to hear Britain's ambassador to Lebanon on the other end of the line offering him a business opportunity. Peter Wakefield, Britain's envoy, was a fluent Arabic speaker who had seen coups and unrest across the region throughout his diplomatic career. However, in recent months, the situation in Beirut had become too dangerous even for him to handle, and most of the embassy team had evacuated to London. Western diplomats were a precious commodity in the conflict, and British embassy staff were worried they could be kidnapped or killed. Now, Wakefield was plotting his comeback, and needed a little help from KMS.

Barely a week after the phone call, Wakefield was sitting on Middle East Airlines Flight 202. It was the first flight into Beirut since the airline had grounded its planes over safety fears. The flight was packed with high-profile passengers – from the airline's chairman to a BBC reporter. But Wakefield was taking no chances. He was flanked by a 44-year-old armed bodyguard, Robert Extance Creighton from KMS Ltd.[30] It was a wise precaution. Creighton was an old colonial – born in Hong Kong, he had served in the SAS with distinction in the jungles of Malaysia.[31] For the next month, he shadowed the British ambassador in Lebanon, living with him in his residence and constantly shielding him from harm. Although such a VIP service did not come cheap, Wakefield was 'well content' with the KMS man. When the pair flew back to London together shortly before Christmas 1976, Creighton was met at the airport by his wife, a young ex-policewoman. The scene at arrivals was more than just a romantic reunion. It was a chance for the ambassador to vet the bodyguard's wife.

Wakefield would return to Lebanon in 1977, and this time his visit would not cause the Creighton couple to live apart. The British embassy in Beirut had found a vacancy for Mrs Creighton as a receptionist, controlling access to sensitive parts of the building, and operating a radio network between the ambassador's residence and his car. When she was not at work, she could live with her husband in the ground floor of the residence. The ambassador's concern for his bodyguard's marriage was touching, and he told the Foreign Office he hoped the pair would now be more 'settled'. The KMS contract in Beirut reflected the company's relationship with Whitehall in the mid-1970s. A small, intimate, almost family affair.

This cocoon was nearly shattered at the end of 1977, when the dangers of Beirut became painfully apparent. A car crash killed two members of the British embassy team in Beirut. A KMS man in the convoy, Ken Ovenden, had a lucky escape. 'There is nothing he could have done to prevent the accident', Ambassador Wakefield wrote in a letter to KMS, 'but the calm and effective way in which he dealt with the situation, summoning the police, personally recovering the bodies, and later giving us a highly detailed account of the accident, was highly commendable and an example to us all.' The bodyguard had conducted himself in an 'exemplary manner', and the ambassador's praise 'gave great pleasure at KMS'. Once again, the company's feelings were paramount. 'Knowing that they are appreciated helps keep them ready to extend themselves for Beirut or elsewhere at short notice when needed', wrote the Foreign Office's head of security, Chris Howells.[32] The company's dedication to British diplomats in Beirut would earn it more than just praise. By 1980, the Lebanon contract was earning KMS £13,335 each quarter, which would now be worth nearly a quarter of a million a year. The bodyguard team had tripled since Creighton first arrived.[33] The deal, originally signed by a Labour government, had proved to be a nice little earner.

* * *

The fact that Britain's Foreign Office awarded work to KMS in 1976 was remarkably good fortune for the firm. It was an otherwise torrid year for the nascent private security industry. Around 160 British mercenaries had flocked to Angola, where Portuguese colonialism had only just retreated, leaving behind a civil war between rival anti-colonial factions. Angola's Marxist movement was by far the strongest, bolstered by Soviet-supplied tanks and trained by Cuban military advisers. Their claim to authority was contested by several Angolan opposition groups, backed by the US and an assortment of Western mercenaries.

These hired guns were an eclectic bunch of young men, with a mixture of combat experience. Some were die-hard anti-communists, others say they simply wanted to earn a decent wage to afford a family home. Their recruitment, often from London pubs, was fairly brazen – little attempt was made to keep the plan secret. John Banks, an Englishman who hired most of the mercenaries, claimed that he even told the Metropolitan Police's Special Branch detectives what he was doing.[34] TV cameras at London airports captured the gung-ho men streaming on to Angola-bound flights. Money for their efforts was readily available, although its exact origins were unclear.

KMS appeared to have limited involvement with this operation, although *Time Out* magazine claimed that two of the company's managers were 'acquainted' with chief recruiter John Banks.[35] The Conservative peer, Lord Carrington, would later grumble privately about 'the past connection of that company [KMS] with the provision of mercenaries in Angola and elsewhere'.[36] Yet KMS, with its discreet Foreign Office bodyguard contract in Beirut, was regarded by many as the polar opposite of the Angola mercenaries. 'Other companies of that nature didn't exist at the time', one KMS veteran, Robin Horsfall, told me. 'The only previous business models had been recruitment of real mercenaries for operations like the Congo and Angola, which were riddled with all kinds of corruption and problems.' He regarded them as 'a bunch of bandits'. By contrast, KMS was a 'proper legitimate company' with 'strong moral regulations and military regulations'.

It is not hard to see why some KMS staff would balk at the comparison with the mercenaries who fought in Angola. Few of them had the same special forces experience as KMS – some were paratroopers, but others were just teenagers with no combat experience whatsoever. The operation descended into an international debacle in February 1976. It was led by a Greek Cypriot with jet black hair, Costas Georgiou, who had previously served with the British army in Northern Ireland as a member of 1st Battalion Parachute Regiment, when it was involved in the notorious massacre of 14 Catholic civil rights protesters on Bloody Sunday – an incident in which some believe he personally fired dozens of rounds.[37] A few months later, he stole weapons from the regiment's armoury and robbed a Post Office, landing himself in jail. On his release, he reinvented himself as 'Colonel Callan' and sought his fortune in Angola. When the fighting failed to go his way, he ordered the massacre of 14 mercenaries under his command who wanted to quit.

This cold-blooded murder in the remote Angolan town of Maquela do Zombo had ramifications that reverberated through Westminster. A British mercenary commander had executed his own men. Although KMS was not involved, the scandal would almost close down the company. It certainly would have looked that way at Courtfield Mews when Prime Minister Harold Wilson addressed Parliament from the green benches. 'The whole House will, I think, have been as disturbed as I was by the evident facility and speed with which a small group of people, funded by an unknown source, were able to recruit misguided people to participate in ... this bloody business', Wilson said sternly.[38] 'The potential dangers of such easy recruitment are apparent, but the proper form of control is not easily defined – and the existing law on the many complex issues involved is unsatisfactory.' The prime minister announced that he was setting up an inquiry to explore whether there were sufficient controls over the recruitment of UK citizens as mercenaries. The inquiry was directed to consider the need for legislation, including possibly amending the Foreign Enlistment Act. This archaic law, passed in 1870, was the only one on the statute books dealing with mercenaries.[39] It was antiquated and had very narrow parameters. Wilson

said scornfully, 'One has only to read what it says about princi-
palities, Powers, peshwas and all the rest of it.' It banned British
subjects from 'any commission or engagement in the military or
naval service of any foreign state at war with any foreign state at
peace with Her Majesty' – but no one had been convicted of this
offence since the Act was passed over a century ago. The inquiry,
known as the Diplock committee, was designed to 'prepare
the ground for any necessary changes in the law' and asked to
proceed 'with all possible speed'. It looked like Keenie Meenie's
days were numbered.

* * *

Despite the debacle in Angola, the private security industry still
had some allies in Parliament. The leader of the opposition,
Margaret Thatcher, was hesitant about the inquiry. She expressed
a degree of caution, warning that 'British citizens have within
present recollection fought for many different causes overseas'.
The 1973 Arab-Israeli war was still fresh in her mind, where a
number of Thatcher's Jewish constituents from Finchley went
to fight for Israel. 'That, I am sure, is understood by everyone',
Wilson reassured her. 'Similar considerations arise in other parts
of the world. That is quite different from the present situation. It
must be a matter for the Committee to look into.'

The prime minister saved his ire for the mercenaries in
Angola. He was alarmed that 'small time crooks with records
have become possessed of vast sums of money, sums far greater
than they could ever earn in other ways honestly or dishonestly,
and have obtained access to lists of names of former soldiers, SAS
and the rest, and signed them up as mercenaries'. Wilson was
hinting at something deeper. He was not solely concerned with
the guns for hire, but with whoever was doing the hiring. 'That
this is possible could be a threat to democracy in this country',
he warned. 'They have raised a vast private army because they
had access to money to enable them to do so.' Wilson's fear

prompted some MPs to sneer, but the prime minister was clearly concerned about where the money and easy access to veterans had come from.

Many Labour MPs, who were growing tired of British colonialism, supported Wilson's efforts to set up an inquiry, but criticised him for doing nothing to stop mercenaries being recruited immediately. 'Would it not be possible to put an Act through this House in 24 hours, at least making it illegal to openly recruit mercenaries in this country?' one asked.[40] Others worried that banning Britons from fighting abroad would be forgetting 'the courageous contribution made to the Spanish Republican cause by some of our own colleagues', referring to the British left-wingers who fought against Franco's dictatorship in the 1930s. Ironically, this view was shared at both ends of the political spectrum. Julian Amery, a hawkish Tory MP from a right-wing family (his brother was hanged for disseminating Nazi wartime propaganda), was determined to speak in favour of mercenaries. As a cabinet minister in the 1960s, he was one of a handful of leading Tory politicians who had incited former SAS men such as Colonel Johnson to fight Nasser's forces in Yemen. Still committed to prolonging the last vestiges of European colonialism, he told the House: 'Many of us feel that the Western Powers collectively should be organising help to the pro-Western and anti-Communist forces in Angola and that, in the absence of such help, to interfere with the flow of genuine volunteers to the pro-Western forces in Angola would be tantamount to becoming accomplices of the Cuban and Soviet aggressors.' This debate over mercenaries reflected, at that moment in the mid-1970s, a wider culture clash about the future of post-colonial British identity. It was a period of intense political flux, before the ascendency of ideologically assured leaders such as Thatcher and Reagan. In 1976, Parliament, like the country, was torn between those who wanted to cling on to empire, led by crypto-fascist Monday Club members like Amery, versus the liberals and socialists who wanted to wash their hands of it. In this context, a ban on mercenaries would have been seen as a defeat for empire itself.

*   *   *

Amery need not have worried. Wilson had picked a right-wing judge, Lord Diplock, to chair the 'Committee of Privy Counsellors on the Recruitment of Mercenaries'. Diplock was best known for introducing trials without jury for terrorism cases in Northern Ireland, earning him the wrath of the Irish Republican Army (IRA). He was certainly creative when it came to the law. First, Diplock canvassed civil servants within Whitehall for their views on mercenaries. Many had a pragmatic, almost Machiavellian approach. A Home Office securocrat said: 'it seems to us that the case for controlling mercenaries rests almost entirely on considerations of foreign policy'.[41] Within the Foreign Office, whose lavish Durbar Courtyard is furnished with statues of mercenaries from the East India Company, diplomats were almost unanimously opposed to more legislation. One asked how Britain could ban mercenaries, when the army routinely hired Gurkhas from Nepal. And a ban could complicate a key lever of British diplomacy – sending police and soldiers abroad to train foreign forces, 'all of which we positively encourage'.[42] It quickly became apparent that the concept of a mercenary could be very broadly interpreted to include a range of UK government employees.

Even when the mercenaries were not employed by the Crown, sometimes they offered 'advantages' to British foreign policy. In practice, the UK could participate in a war unofficially by turning a blind eye and allowing British soldiers of fortune to go out and fight. If another government complained, one diplomat suggested, Whitehall would only need to issue 'disclaimers and cosmetic action' about not being directly involved with those mercenaries to prevent any damage to international relations. (This would prove to be prescient, as a major trading partner, India, did later object strongly to the work of KMS in Sri Lanka, and Whitehall deployed such a disclaimer.) There was also debate as to whether a distinction might be drawn between 'recruitment for gain and recruitment for political sympathy', anticipating that some Britons might soon want to support white Rhodesian relatives in their colonial struggle against granting black majority rule.

Two months after the Diplock Committee began, and having imbibed these reactionary views, it was poised to present its report to the prime minister (now Labour's Jim Callaghan, who had taken over from the ailing Harold Wilson). However, events in Angola gave Diplock reason to withhold publishing his inquiry. Colonel Callan and a dozen of his men had been captured by Angola's Marxist government, the Popular Movement for the Liberation of Angola. The streets of Luanda thronged with demonstrations against 'mercenarism', and the foreigners (including ten Brits) were due to be put on trial. Diplock did not want his report to prejudice the hearing. He planned to draw a distinction between 'regular armed forces of sovereign states, whose members are entitled to the status of belligerents in international law' (i.e. captives should be treated as prisoners of war), versus 'private armies and guerrillas who do not benefit from that status and may be treated and punished as ordinary criminals by the state in whose territory they operate'.

Diplock fretted that 'anything we said upon this subject might be misused to justify the denial of British mercenaries of the status of prisoners of war and the infliction on them severe or even a capital penalty'. He suggested to the prime minister that he delay submission of his report 'until the possible danger to the mercenaries is over'.[43] The prime minister obliged. As much as Downing Street wanted to appear tough on mercenaries, they did not want to be blamed for getting any of them killed. The trial was set for June, and Diplock could publish his report after the verdict. The Luanda hearing was televised in numerous European languages, turning it into a Western media spectacle, climaxing with Callan's death sentence. Angolan officials told visiting journalists that Callan was a 'racist with a fascist mentality, and a man with extreme contempt for life'. His former comrades had little sympathy, describing him as a 'homicidal maniac, who spent a lot of time killing blacks just for fun'. Four of the mercenaries, including Callan, were executed by firing squad, the remainder sentenced to lengthy jail terms. It was a powerful sign that former colonies would no longer tolerate European mercenaries trying to topple their new independent governments.

As the dust settled, Diplock finally presented his 27-page report to Britain's prime minister. He started almost philosophically: 'The soldier of conscience may be found fighting side by side with the soldier of fortune', Diplock wrote lucidly. 'The motives which induce a particular individual to serve as a mercenary will be mixed. A spirit of adventure, an ex-soldier's difficulty in adjusting to civilian life, unemployment, domestic troubles, ideals, fanaticism, greed, all may play some part.' He reasoned that mercenaries should therefore be defined on the basis of what they do, and not why they do it. On this basis then, a mercenary was simply 'any person who serves voluntarily and for pay in some armed force other than that of Her Majesty'.

Diplock agreed that the existing legislation, from 1870, was so antiquated that it was riddled with loopholes. For example, a mercenary who left the UK by plane, rather than ship, would not be liable to prosecution. The draftsmen of yesteryear had clearly not anticipated the advent of air travel. Diplock noted dryly that a ship is 'a means of travel unlikely to be used by a mercenary in modern time'. He was also concerned about the insurmountable difficulties of bringing mercenaries to trial, as it would depend largely on testimony from comrades, who could not be compelled to incriminate themselves.[44] The judge noted that the director of public prosecutions had decided not to prosecute any British volunteers in the Spanish Civil War because of these evidential difficulties.

The old Act was so ineffective that Diplock suggested most of it should be repealed and a fresh start made. But he was surprisingly permissive. 'Enlistment as a mercenary by a UK citizen abroad should cease to be a criminal offence', he said. His concerns stemmed from an extreme libertarian standpoint, in which to 'prevent a UK citizen from accepting service as a mercenary abroad is a restriction upon his personal freedom which could only be justified on ground of public interest'. There were also practical concerns about international relations: 'We do not think it can be justified on grounds of public interest to impose a general prohibition on United Kingdom citizens from serving in some capacity or another (e.g. as instructor or technician) in the armed forces of a friendly state at a time when there are no hostilities in which that force is engaged. To make it a criminal offence under UK law for him not to desert that force as soon as it became involved in external or internal conflict, to which the UK was not itself a party, would, we think, be an impermissible affront to the sovereignty of the foreign state concerned.'

These objections aside, Diplock turned his focus to companies in the UK that recruited mercenaries, and proposed that it was them who should be the target of any fresh legislation. Companies that offered employment as a mercenary, facilitated payments to mercenaries or even just published information about how to find work as a mercenary, should be banned. If approved, this would mark the end of companies like KMS. However, Diplock was conscious that the government did not always disapprove of mercenary activities, and suggested an important caveat. The power should only be exercised in circumstances where the recruitment of mercenaries in the UK would have an adverse effect on international relations, for instance if the company were fighting against a British ally. As such, it should be left to ministers to decide which mercenary operations were undesirable on a conflict-by-conflict basis.

The prime minister thanked Diplock for his report, and the home secretary told the Commons that its recommendations would be considered. Behind the polite veneer in Parliament, Whitehall went to work quashing its recommendations. Civil servants immediately noted that Diplock's definition of a mercenary would cover UK contractors and loan service personnel serving in Oman. This suggests officials were conscious of safeguarding the KMS deal with the Sultan's Special Force, or the myriad other British troops on loan to Gulf regimes. The Foreign Office's West Africa Department went further, and asked 'whether we need any legislation at all'. They reasoned that not having any laws against mercenaries would allow the government to 'shrug our shoulders and say we had no power to stop it'.[45] The department went on to warn, in what appears to be another thinly veiled reference to KMS, that 'There are a number of "respectable" British mercenaries active e.g. in Oman, some recruited by HMG [Her Majesty's Government]. In other similar circumstances they have enlisted direct but with our blessing. Would we for example wish to stop the Embassy of a friendly government recruiting British mercenary pilots for their air force?' Lastly, if ministers were required to specify which conflicts were prohibited, they asked, 'What would our attitude be in the event of a rebellion against an oppressive or hard-line communist regime?' Ultimately, the diplomats stressed that any legislation against mercenaries should 'give enough flexibility'.

This view was shared by many in the Foreign Office. An official in its Middle East Department stressed: 'Contract personnel, by definition mercenaries, did a valuable job in the Middle East and enjoyed a measure of official approval.' He went on to argue that the Foreign Office should highlight the disadvantages of new legislation when speaking to ministers. The civil servants roundly agreed that ultimately 'it was best to have no legislation'.[46] The department's orientalism was plainly on display, commenting: 'Mercenaries are seen, in Africa in particular, in emotional terms, as a manifestation of continued white colonialist interference in affairs that no longer concern them.'[47] Others pointed out that the UK's own Ministry of Defence was

White Sultan of Oman

involved in helping recruit ex-servicemen for foreign armies, even vetting applications for those who wished to serve in the army of apartheid South Africa.[48]

As the debate rumbled on into the winter, with diplomats determined to drag their feet, another concern cropped up. The Middle East Department reiterated that 'our firm preference is still for no legislation at all ... we should present to them [ministers] the benefits of doing nothing'. The department also warned that certain recruitment agencies had recently come to its attention. By now it was December 1976, and a KMS bodyguard had spent most of the last month protecting Britain's ambassador in Beirut. 'One, KMS, is known to us', the diplomat said, 'and has substantial business with a number of Middle East Governments.' He warned that Diplock's proposed legislation could put Downing Street in the awkward position of having to approve or disapprove of the actions of KMS, which could cause embarrassment to the company's clients.[49] The Security Department chimed in the next day, saying that KMS Ltd 'are a reputable firm in their line of business, employed for example by Sheikh Yamani (Saudi Arabia's oil minister) as probably the best and certainly most reliable contract bodyguards in the world'. (The company would also appear to have won work with governments in the United Arab Emirates and Zambia.)[50]

A distinction was being drawn between KMS, due to their direct involvement with the Foreign Office, and the unruly mercenaries in Angola that gave the private security industry such a bad reputation. The civil servant gushed:

I value the FCO [Foreign and Commonwealth Office] contract with them [KMS] which provides high grade professional armed protection at short notice for Ambassadors under threat. At the moment the Ambassador in Beirut and the Charge D'Affairs in Buenos Aires are protected by KMS bodyguards. We might have to call on their services elsewhere at any time, and can always do so at short notice. If KMS Ltd were legislated away *en passant* no comparable substitute protection would be available to the Diplomatic Service.[51]

These concerns were contrary to the mood among Labour MPs, who had proposed a short bill to ban the recruitment of mercenaries in their legislative programme for 1977. Still, almost six months after Diplock submitted his report, diplomats were debating his proposals to ban mercenary recruitment. Not only would Diplock's simple definition cause problems for KMS, it could also jeopardise loan service personnel, members of the British military seconded to foreign forces, as was common in the Gulf. 'The provision of loan service personnel is an important element in our political relations with a number of countries in the Gulf region', a diplomat explained. 'It also contributes to the security of an area from which we import over 80 per cent of our oil.' There were 225 such personnel serving with the Sultan of Oman's armed forces in 1977, and Diplock's definition would encompass all of them. The concerns went on: 'If, in order to exclude certain governments such as the South African from being permitted to recruit, we were to introduce into legislation some kind of licensing policy, this would also provide an opportunity for opponents of our policy in countries such as Oman to press for their exclusion also.'[52] After months of internal wrangling, and shortly before breaking for Christmas, the diplomats reluctantly agreed to propose to ministers a ban on recruiting mercenaries except by UK government agencies or her allies. But Harold Wilson's replacement, Callaghan, seemed to lack his predecessor's enthusiasm for a crackdown on mercenaries. The proposed legislation appeared to die a death, much to the relief of the old brigadier at Courtfield Mews.

2

Bodyguards and Business Building

Ireland, 1976

It was ten past two in the morning. The car, a Hillman Avenger, was frantically weaving through the narrow pitch-black country roads. Malcolm had been up all night. The car collected him hours later than planned. It was not even the right car. He had expected a Triumph 2000. Where on earth that vehicle was now, he had no idea. Then suddenly a roadblock appeared up ahead, near a wooded glade. As the figures came into focus and hailed the car to halt, it became clear they were not fellow British soldiers. It was the Gardai, the Irish police. This was the last thing he needed. Staff sergeant Malcolm Rees, a wiry ginger-haired SAS man, was sitting in the car dressed as a British paratrooper. He was not meant to be seen south of the Irish border. And here he was, half a kilometre inside the Republic of Ireland. The driver had taken them south, running parallel to the Newry River. The car was full of weapons. Their arsenal included a sawn-off shotgun, albeit an official British army-issue Browning, fitted with a 20" barrel to make it less cumbersome to carry in a car.[1]

Rees was not used to being lost. He was an expert cross-country runner, and as a young soldier in the Royal Signals Corps he easily outpaced his unfit superiors on the obligatory mountain runs. With a penchant for adventure, he served with the SAS in Operation Storm, a major offensive against the revolutionaries in Dhufar that had ultimately emerged from the 'Landon Plan'. He was a communications expert, and his commander praised his work there as 'outstanding'. Especially during the last ten weeks of his tour in Oman, he kept the vital radios running while under fire every day from small arms, machine guns and mortars. When one of his sergeants was fatally wounded, Rees had a 'calm and unemotional appreciation of the situation', which allowed him to 'steady and direct his comrades'. He won the Military Medal, a rare honour, for his 'coolness and determination under fire'.[2]

Five years later and Rees was sitting at this Irish police checkpoint trying to figure out an exit strategy. Suddenly, a Vauxhall Victor pulled up behind them. Here were more of his men. Now there were six British soldiers, heavily armed but mostly wearing civilian clothes, caught driving on the wrong side of the border in the middle of the night. Even their army identity cards did not satisfy the police, who suspected they might be criminals or worse – terrorists. The irony was that his job that night had involved lying completely still for hours in a hidden observation post near a primary school, hundreds of metres from the Irish border. He was on the look out for IRA men crossing into South Armagh, one of the most dangerous parts of Northern Ireland. Perhaps he had wandered onto the wrong side of the border. So what? The IRA paid no attention to it. Why should he? Why should the Gardai? There was no point reasoning with them now though. He ordered his men to surrender. He did not want anyone getting shot. His superiors would sort it out and it would all blow over quickly. It was just a misunderstanding, surely? Why were the Irish police being so uptight anyway?

The Gardai took Rees and his men to Dundalk police station. Inside its thick stone walls were two more of their comrades, the missing lads from the Triumph 2000 who were originally meant to pick up Rees from the lookout point. They too had been stopped on the wrong side of the border. Neither of them were wearing uniforms, and they had a submachine gun and pistol between them. In total the Gardai had found over 200 bullets on them. Perhaps this would be harder to explain. Within hours, their arrest was causing chaos at the Northern Ireland Office, with civil servants panicking about the story leaking and causing 'intense criticism'.[3]

* * *

The SAS had arrived on the other side of the border, in South Armagh, several months earlier. Their deployment was codenamed Operation Contravene and followed a massacre near the village of Kingsmill of ten protestant workmen on their bus ride home. Unusually for this secretive British army unit, Prime Minister Harold Wilson had announced the arrival of the SAS, perhaps hoping the regiment's fearsome reputation would cause the local IRA men to back off. Battle-hardened members of D Squadron SAS came straight from mopping up the Dhufar revolution in Oman. They took a while to acclimatise to their new surroundings. The 'yellow card' rules were a far cry from the free-fire zones in the *jebel*. In Northern Ireland, they could only open fire if someone posed an immediate threat to life. Still, their old tactics of covert surveillance, intelligence gathering and ambushes could work well here. They had proved themselves in mid-April, when they snatched an IRA staff officer, Peter Cleary, from his fiancée's house at Forkhill, less than a mile north of the border. The 25-year-old motor mechanic had been under intense military surveillance for days, as the SAS waited for him to visit Shirley, his wife-to-be. A fortnight before their wedding, on 15 April, soldiers spotted Cleary tinkering with a car outside her house. As a moonless night descended, a neighbour's dog started to bark. A local resident shone a torch across his front garden and saw the blackened face of a man in battledress. A figure wearing a red beret stood up and fired a warning shot.

'We have you surrounded', he yelled, as his colleague leapt over the garden wall and steamed towards the house. Shirley was sitting in the living room watching the *News at Ten* when a tall, well-built soldier with dark hair and blackened face burst through the front door wielding a rifle. 'Everyone outside', he barked. Another soldier appeared, brandishing a revolver. Peter was led out first, and everyone was lined up against the front of the house. The soldiers violently frisked the menfolk. Shirley saw her fiancée being pushed into a pile of sand and laying face down on the floor as a soldier pointed a gun at the back of his neck. She tried to run over to him, but a soldier pushed her away and yelled 'Get up against the wall'. She grabbed hold of her sister's seven-year-old daughter, who was crying, much to the consternation of one of the soldiers who allegedly warned 'Shut up or I'll shoot you'. After 15 minutes, Peter was led away from her side by two soldiers. Everyone else was ordered back inside the house, as the soldiers smashed an exterior light and pulled out electricity wires. 'You are getting married shortly aren't you?' A soldier taunted Shirley. 'I have news for you, you aren't going to be', he laughed. As they left, the family were warned, 'If any of you put your heads outside the door for the next 15 minutes you will get it blown off!' Terrified, they lay flat on the floor.

After five minutes, Shirley could no longer bear it. She summoned up the courage to go outside and look for her fiancée, but her relatives stopped her from leaving. Then three gunshots rang out into the night from somewhere in the distance. Panic descended as the deafening noise of a Puma helicopter came overhead. The family did not know it then, but Peter had just been killed. His corpse was lying face down and arms outstretched, 600 metres from the house, in a field surrounded by a stone wall covered in ivy. Some of the soldiers at the helicopter landing site were also confused by the gunfire. They had not heard any warnings before the shots were fired, and thought they were under attack. The next thing the SAS troops saw were two of their colleagues dragging Cleary's limp body by his armpits towards the helicopter, smearing the field with a trail of blood over 20 metres long.

Cleary's corpse was flown to Bessbrook barracks, the SAS base in South Armagh, and wrapped in a plastic sheet. A detective sergeant from the local police, the Royal Ulster Constabulary (RUC), arrived there in the morning. He asked to interview Captain 'A', the SAS patrol member who shot Cleary. An army lawyer told him this was not possible at that time. Undeterred, the police returned twelve hours later to meet Captain 'A', the army lawyer and Major Brian Baty, the commanding officer of D Squadron in South Armagh. Baty was a formidable operator, known as 'The Baron'. He had joined the army as a squaddie, before eventually rising through the ranks to become an officer. He had served extensively with the Argyll and Sutherland Highlanders, a tough Scottish regiment that saw action against the Indonesian army and in the Middle East. In Aden, Yemen, he had served under a ruthless officer called 'Mad Mitch', who put Baty in charge of intelligence. In the British army's efforts to crush the anti-colonial revolt in Aden, enemy prisoners were allegedly tortured.

With Baty by his side, Captain 'A' stayed silent as the police cautioned him. The army lawyer then asked to speak to his client alone for 15 minutes. When they returned, the captain had been advised not to give a written statement, but he was prepared to answer oral questions from the police. The interview finished after an hour and a half. Captain 'A' subsequently gave a blood sample and three weeks later supplied a written statement, to be used purely as a witness statement rather than as the accused.

The SAS were making a name for themselves in South Armagh. Cleary's family said the SAS had executed him in cold blood. The army said Cleary was linked to the massacre at Kingsmill, the very reason the SAS had been deployed. More to the point, they said Cleary assaulted his SAS guard as the helicopter came in to land. The guard had asked Cleary to stand and face him, but instead Cleary lunged towards him in an apparent attempt to overpower him and take his weapon. In response, the guard fired three rounds at Cleary from a distance of four feet, hitting him once in the chest or side of the chest.[4] The army felt this statement vindicated the actions of the SAS, although it is hard to understand how Cleary assaulted the guard from four feet away. Whatever the truth, the killing did little to win the hearts and minds of local people, and the Gardai took no chances with their eight SAS captives. They were hauled from the Gardai station in Dundalk and taken down to the Special Criminal Court in Dublin. Their heavily guarded convoy was booed and jeered when it left Dundalk, and a crowd outside called them murderers and killers.[5] They sat silently in the dock, as Mr Justice Pringle read out their charges – two counts of possessing weapons and ammunition with intent to endanger life, and having no firearms certificate. Their real names were announced in court and splashed on the front of newspapers, blowing their cover. The British embassy's military attaché in Dublin watched anxiously from the gallery, as all the men were released on £40,000 bail. Over in Westminster, a defence minister told Parliament that the men had merely made a 'map-reading error'. After the bail hearing, the SAS team were flown by chopper straight back to Bessbrook, where they were greeted by Baty. He was unsympathetic and ordered their interrogation by military police. Their comrades protested that they were being treated like common criminals, causing a 'near mutiny'.[6] The director of the SAS and the regiment's commanding officer had to fly to Bessbrook to calm them down – by pledging to resign if the men were put on trial.

The regiment was right to be worried. The incident had caused civil servants to profoundly question their profes- sionalism. In a secret memo, one official warned that it was 'questionable whether units like the SAS should be deployed in an internal security operation at all ... where success depends far less on military derring-do than on at least securing the passive acceptance of the security forces by the local population'. They noted the 'capacity of the SAS to do something to embarrass' the British government, saying that the 'SAS has an unfortunate reputation'. The border crossing and killing of Cleary had 'done nothing to improve that image'. The official felt it was too risky to expand the SAS deployment in Northern Ireland beyond South Armagh, and that even their presence there was little more than a 'gimmick' and 'cosmetic'.[7]

The following March, the SAS men were put on trial in Dublin. The case lasted ten hours over two days. Major Baty gave embarrassing evidence in defence of his supposedly elite men, repeating the claim that the British army's best troops had somehow made a 'map-reading error'. Baty testified that the 'direction to me, and the direction passed by me to the soldiers, was that the border was inviolate, and that we were not to cross the border on foot, by vehicle, or any other means'.[8] He clearly managed to convince the court that it was an innocent mistake, as his men were let off with a £100 fine each. The media coverage was more bruising, with a mocking cartoon on the front page of the *Guardian*, showing soldiers learning to read a map. The publicity was more than a source of ridicule. It was a rare window into the SAS command structure, revealing Major Brian Baty as a key player in South Armagh. His card was now marked. A chain of events had been unleashed that would one day culminate with Baty leaving the special forces and effectively fleeing the country.

* * *

David Walker was a grammar school boy from Southampton. Intellectually able, he studied engineering at Cambridge and joined the British army's Royal Engineers as an officer. He enrolled at Sandhurst in the cohort below Tim Landon and Qaboos, and for a year the trio would have overlapped at the military academy.[9] Walker flourished in the army, and a decade on from Sandhurst he followed in Landon's footsteps – to the revolutionary highlands of Dhufar. By 1972, he was a captain and troop commander in 'G Squadron' – the main SAS unit then in Dhufar. Malcolm Rees had served with the same squadron six months earlier, when he came under heavy fire and won the

Military Medal. Now it was Walker's chance to demonstrate what he was capable of on the battlefield.

Operation Storm, the major offensive against the Dhufari revolutionaries, was still rumbling on. In fact, it was about to reach a climax. Around sunrise on 19 July 1972, a small team of SAS men came under heavy fire from hundreds of enemy fighters. They were being shot at from the hills and pinned down on the roof of a small coastal mud fort at Mirbat. The nine SAS men were vastly outnumbered, and running low on ammunition. They had to run through a hail of bullets to reach their heavier weapons, which had the range and firepower to suppress the rebel advance. Several of the men at Mirbat died that day, and their bravery became the stuff of SAS legend, reconstructed for cinema as a David versus Goliath battle and exhaustively recalled in memoirs. One adoring book about the battle is subtitled *The SAS Under Siege, Nine Men Against Four Hundred*, and inside the authors gush breathlessly about the 'greatest acts of heroism in the long and glittering history of the SAS'.[10]

In reality, those soldiers in Mirbat owed their survival not just to their bravery, but also to air cover from British pilots, SAS ground reinforcements and local allies. In all the myth making, it has largely been forgotten that among those who came to the rescue that day was David Walker, who played a key role in winning the Battle of Mirbat. As the firefight raged down at the fort, a helicopter dropped Walker off near a small ridge, with 15 men under his command. As soon as they landed, they shot dead five enemy fighters who had appeared in front of them. Walker then went over to collect their bodies and weapons, before coming under heavy fire from the fort down at Mirbat where the enemy were beginning to retreat. Walker, a skilled tactician, carefully laid a trap. As the enemy tried to back away from the fort, they became cornered between Walker's ridge, the town at Mirbat and the sea. A group of 30 Dhufari revolutionaries had no way out, and they were 'completely destroyed by Walker and his small group'. The head of the SAS, Peter de la

Billière, commented that Walker's 'calmness, quick thinking and personal bravery played a vital part in the victory'.

Walker could therefore take credit for killing almost half of the Dhufari rebels who died in the Battle of Mirbat. The British military is fond of saying that this battle marked a turning point in the Sultan's struggle against subversion, and the beginning of the end for the Dhufar revolution. However, an Omani contact who has spoken to local people in Mirbat says the incident has been exaggerated. It was not such a significant setback for the rebels, who continued to prosper until the Shah of Iran came to the Sultan's aid with tanks and armoured vehicles in 1974. This source also questions whether the SAS was really so outnumbered at Mirbat. They were actually supplemented by 30 or 40 local troops, who also fought tooth and nail to survive, and there were half as many revolutionary fighters in the field as the SAS makes out. It was more like 200 rebels against 50 supporters of the Sultan, not 400 against nine. Even more damning is the claim the battle did not last for one day, but took place over several days. It began with Dhufari women revolutionaries slipping into Mirbat unnoticed, linking up with local sympathisers and booby trapping the town with mines and ropes. They agree that it was asymmetric warfare – an unfair battle – but that the revolutionaries were the underdogs, bereft of air power and compelled to improvise.

Months later, Walker saw action again, in a daring night raid on Jebel Aram, a lush mountain hideout 500 metres above sea level, mid-way between Mirbat and Salalah. (Military records suggest Malcolm Rees fought alongside Walker in this firefight.) His commander was so impressed with Walker that he said 'his example and initiative ensured the success of whatever group he was with'. He was recommended for the Military Cross, a coveted award. Remarkably though, he never received it. The British army was keen to keep the SAS battle at Mirbat a secret – officially they were only in Oman to train local troops, rather than implement the ambitious goals of the Landon Plan. Instead,

Walker was given a vague mention in dispatches nearly two years later, which made no reference to the SAS. He was described by the Queen as an ordinary Royal Engineer, who had given 'gallant and distinguished service in Oman' at some unspecified period of time. Walker was not the only SAS warrior treated like this. Decades later, SAS veterans banded together to campaign for their Mirbat medals.

Perhaps it was disgruntlement at not being properly decorated in Dhufar that led Walker to quit the army a few years later, at the height of his career. He retired by the end of 1974, and became a reserve member of the SAS who could be called on in times of emergency. Another reason for his departure may have been the pull of the private security industry. He is said to have become involved with Control Risks, a firm that specialised in protecting VIPs from kidnappers. These skills certainly made him desirable to one of the most prestigious names in private security at that time, KMS Ltd. In the early summer of 1976, Walker's name begins to crop up in Foreign Office telegrams. He was still a reserve SAS officer, recently promoted to major, but by now he was a successful businessman, a co-founder of KMS. He was carefully managing one of the firm's first contracts, to supply bodyguards for British diplomats in Buenos Aires.

Walker jetted into Argentina to find a country in turmoil.[11] The military had just staged a coup, and were cracking down on left-wing dissidents with harrowing ferocity. In turn, armed opposition groups were targeting the regime and its foreign investors, including British interests. Despite what the local *chargé d'affaires* tried to tell Walker, there was a real need for bodyguards. The security situation had not improved since KMS first signed the contract, in the wake of that awful day in April 1975.[12] A 10 kilo car bomb had been planted in a Ford Falcon across the road from the British ambassador's house. It was timed to explode during a Queen's birthday reception, but thankfully the timing device had failed. The car did not move for three days, until a local policeman became suspicious and tried to investigate. It turned out the bomb was booby-trapped. If the timer failed, attempts to tamper with it would trigger the detonator. The policeman never stood a chance. The blast was so powerful it blew the boot clean off, smashing all the ambassador's windows and killing several local people.

Added to this volatile mix was Britain's occupation of *Las Malvinas*, or the Falklands Islands, which many Argentinians of various political hues regarded as their own territory. In short, there were myriad reasons why British diplomats in Buenos Aires might meet a grisly end. The Foreign Office trusted Walker and KMS to keep their men safe. The price, £70,000 per annum (now worth nearly half a million pounds), was not a problem.[13] Walker later recalled at a colleague's funeral how the costs of this deal were calculated on the back of an envelope and, 'in true entrepreneurial spirit the formal legal contract was signed on behalf of a company which did not yet exist'.[14] The nascent firm had an important ally in the shape of Chris Howells, head of the Foreign Office's Security Department. He came straight from a stint in Warsaw, where the KGB was constantly probing the British embassy, and did not take any chances when it came to the security of his fellow diplomats. He told the sceptical *chargé d'affaires* that there were other factors to consider as well. 'It is a shade callous', Howells said, 'to point out that your guards are provided mainly to save Ministers and others here from the intolerable complications in relations with Argentina that would follow from a mishap to you. The consequences would be smartly linked to the Falklands issue in some way or another.'[15]

Walker made sure the KMS team were on high alert, to such an extent that the British envoy grumbled about his bodyguards being an 'omnipresent entourage' who made him 'the most heavily guarded British head of mission in the world'.[16] Our man in Buenos Aires continued to complain: 'I have no privacy left to me, not even in my own home – except in the lavatory and in bed … I am escorted for 24 hours a day by armed thugs who dog my every footstep.' The worst indignity for this diplomat was when they were 'even condemned to have breakfast together (which is the time of day when I am at my most fragile)'. The whole experience of KMS chaperones had left his 'nerves as taut as banjo strings', and he begged the Foreign Office to pay for him to go on an extra holiday. His leave was granted, but the KMS guards would not be stood down. Howells was boosting their pay by 7.5 per cent, based on a 'cross subsidy from their profitable business in the Middle East'.[17] The deal would now cost the Treasury £80,000 in 1977, and looked set to continue long into the future.[18]

* * *

The Netherlands, 1979

Two middle-aged men wearing dark suits and coats lingered outside a smart house in The Hague. The building was easy to find, with an elaborate archway at the entrance above which stood a weathered Gothic statue. Around 9.00 am, they saw an older man with thick black-rimmed spectacles emerge from a door and walk into a gated compound behind the archway. His Rolls Royce silver shadow was waiting for him, with his chauffeur at the wheel and teenage valet holding the door open. At that moment, the pair fired eight shots. The valet hit the pavement and the old man was propelled into the passenger seat as the chauffeur sped off. The passenger and valet were bleeding from serious head injuries and would die on arrival at hospital. Meanwhile, the gunmen escaped down a quiet alleyway.[19]

For Britain's Foreign Office, it was a shocking assassination of one of its most senior diplomats, Sir Richard Sykes. For the IRA, it was a daring attack on the British establishment abroad, showing their hitmen could operate across Europe in the cause of Irish Republicanism. For KMS, it was another business opportunity. Hijackings, kidnappings and assassinations were becoming alarmingly common, and protecting VIPs was something that its staff, mostly ex-SAS men, were highly experienced at. The year 1979 would see their supply of bodyguards to British embassies around the world grow fourfold, jumping from three to twelve men.[20] Not long after Sykes was gunned down in The Hague, the Treasury paid for two KMS staff to guard the embassy in the Netherlands, earning the company around £10,500 per guard annually (now worth £82,000).[21]

The company's services were also in demand at the British High Commission in Kampala, Uganda. This would also prove to be a fortuitous posting. The country's military dictator, Idi Amin, had finally fallen from power, and the streets were in a state of turmoil. Four KMS bodyguards were sent to protect the British diplomats, and it was not long before Uganda's new president, Yusuf Lule, was worried about his own security. 'President Lule is interested in acquiring a British expert(s) to train a Presidential Protection Unit in Kampala', a Foreign Office staffer reported in a secret telegram. The next two lines are censored, but it goes on to say: 'It is obviously in our interests to help President Lule and to do what we can to maintain security for him and his Government. But the appointment of a British citizen to undertake this task in Kampala could attract press interest' – more censorship follows. 'On balance, I recommend that [redacted] should be authorised to offer help if approached e.g., in steering the Ugandans in the direction of a suitable candidate, but on the strict condition such assistance should remain absolutely confidential and that there should be no public suggestion that the involvement of British personnel in the training of a Ugandan Presidential Protection Unit had been arranged or endorsed by the British Government.'[22] Nearly four decades after the telegram was written, the Foreign Office rejected a freedom of information request for an unredacted copy of this telegram. They said that the censored content was 'supplied by, or relating to, bodies dealing with security matters', such as Britain's intelligence agencies. Fortunately, despite the best efforts of the censors, they failed to redact a diplomat's handwritten note at the bottom of the telex, which said 'I agree [with the recommendation]. We might discuss at this afternoon's meeting. There are obvious pitfalls, and advantages. The latter outweigh the former.' The very next day, another colleague wrote at the bottom of the telegram in almost illegible script: 'We discussed. The question is being handled by KMS without any official involvement by us. There is no action we need to take at present.'[23] This episode shows, despite the attempts at redacting the telegram, how KMS enjoyed a privileged relationship with the UK Foreign Office. It was not an ordinary security company. Its staff, all former or even reserve SAS men, could be relied upon to handle situations where the UK government wanted to deny any involvement. If Whitehall had officially provided bodyguards to Uganda's new president, it might have looked like the former colonial master was refusing to let go of the reins, or bolstering a puppet president. With KMS as its proxy, Britain achieved the same level of access to an African head of state without appearing as the imperial overlord.

* * *

San Salvador, 1979

The ramifications of the Sykes assassination rippled around the world. It would see the fortunes of KMS improve as far afield as El Salvador, the smallest, most industrialised and densely populated country in Central America. In 1979, it was on the verge of civil war, after a US-backed army general, Carlos Humberto Romero, took power in elections that were widely held to be fraudulent.[24] Successive Salvadoran regimes since the 1930s had seen the army, large landowners and industrialists maintain their grip on power through ballot rigging and intimidation, provoking a left-wing revolutionary backlash supported by progressive elements in the Catholic Church. A British diplomat described the Romero regime as 'strong right-wing and authoritarian', although it was 'not totally autocratic'.[25]

El Salvador was fast becoming a classic Cold War proxy conflict, where Western diplomats were seen as prime targets for the young radicals. The US ambassador had 18 secret service agents plus US marines to look after him. In contrast the British embassy was only protected by armed plain-clothed local police during office hours. Beyond this, less conventional methods were used. 'Additional cover on a 24-hour basis is provided by vigilantes', essentially local muscle who were paid by the British diplomats. This was not an anomaly. The ambassador's residence was 'protected by two armed, uniformed police round the clock with additional cover by vigilantes'.[26]

Despite the risks, Whitehall wanted to keep some diplomatic relations with El Salvador, not least because British companies like Unilever, Shell, Glaxo, British American Tobacco and the Bank of London and South America had investments in El Salvador that may have exceeded £25 milllion.[27] Such corporate interests were heightened when a left-wing armed group, the FARN (Armed Forces of National Resistance), kidnapped two British bankers in the city. When coupled with recent events in the Netherlands, the Foreign Office had decided to reconsider the safety of its diplomats around the world. 'In the light of Richard Sykes' tragic death', an official wrote, 'I am most concerned that all Heads of Mission should be especially alert in protecting themselves against terrorist activity. The latest developments in El Salvador and the threatening statements of FARN make me particularly concerned about your own security.'[28]

The bankers were still being held hostage, and after the Sykes assassination, British diplomats in San Salvador were reminded by Whitehall to make their movements as unpredictable as possible, by constantly varying their route and times of travel. The Foreign Office's own internal security department was struggling to cope with the mounting threats to its diplomats around the world, especially as many of its staff in Europe were now at risk from the IRA. Weeks after the Sykes assassination, the Foreign Office contacted MI5, experts in precautions against kidnappings.[29]

They were tasked with an 'immediate security inspection' of the British embassy in San Salvador, a free-standing two-storey villa in the city centre. A fortnight later, an MI5 officer, Mr Hunter, boarded a British Airways flight to Miami, where he caught a connection to San Salvador.[30]

The British spy spent the next couple of days probing the weaknesses at the embassy, where he found that the ambassador's security was 'at significant risk'. MI5 recommended a slew of security modifications, from an eight-foot-high chain-link fence topped with barbed wire around the ambassador's garden to a bullet-proof car. However, at that stage, an armed bodyguard was considered unnecessary unless there was a specific threat to the ambassador. If such a threat arose, however, then the ambassador should be 'provided at once with a suitably trained and armed UK-bodyguard'. For there to be one bodyguard available at all times would require a team of three.[31]

Shortly after Hunter's visit, a French diplomatic building in San Salvador was occupied by student protesters, and the Swiss envoy was murdered. Whitehall decided to withdraw the British ambassador and his wife until the security modifications could be made and the armoured car arrived.[32] In the interim, his deputy, the *charge d'affairs*, would have to keep a low profile. On 3 November 1979, the situation took a turn for the worse when twelve Salvadoran soldiers held up the embassy's local vigilante at gunpoint in the chancery car park. The army patrol searched the office garage where the vigilantes stored their belongings while on duty, and then 'banged their rifle butts on the door at the rear of the building which lets into the Embassy's backyard'. The vigilante refused to open it and after 20 minutes the soldiers left, having pilfered his pistol and machete.[33]

The handful of remaining junior British diplomats in San Salvador were understandably rattled by the incident, and within days they had accepted Whitehall's suggestion for a team of three KMS bodyguards, all former SAS men, to fly out and protect them. KMS would charge an estimated £18,000–19,000 for the several months' cover (now worth nearly £100,000) and could arrive by mid-November subject to ministerial approval. The KMS guards were told to be discreet and not mention the name of the company that they worked for. Foreign Minister Douglas Hurd authorised their role, albeit with reservations. 'On grounds of principle I am reluctant to see further KMS contracts, but it sounds as if this is inevitable for the moment. So please proceed, but on the assumption that it is for three months only.' The KMS guards were told to live with the diplomatic staff at the ambassador's residence. The job proved to be arduous and soon a fourth KMS guard was requested to reduce the strain of providing a 24-hour watch over the diplomats, although it is not clear if he was ever dispatched.[34]

Within a matter of months, the Foreign Office realised that the security situation in San Salvador was so dangerous that the KMS guards 'would be needed indefinitely', which would have 'effectively doubled the cost of the Embassy'. Whitehall concluded that 'given our limited interests in El Salvador, [it] would have been a disproportionate price to pay to maintain a diplomatic presence'. By mid-February, all British diplomats in El Salvador, including their KMS bodyguards, were withdrawn, having done 'a first class job'.[35] The *charge d'affairs*, who had bravely stayed on in the embassy for many months after the ambassador was first withdrawn, was awarded an OBE for this tenure in El Salvador. Several documents in these files about KMS bodyguards in El Salvador have been censored or removed from the file altogether, suggesting that it remains a delicate topic almost four decades later. The Foreign Office has refused to release the full file, and said the references to MI5 had been released in error.[36] Although the KMS contract in El Salvador was short-lived, it would not be the company's last trip to Central America. In future, the company would be working for a different client in the region with some very bespoke requirements.

* * *

Downing Street, 1979

The year 1979 was not one of change for KMS alone. Britain, as a whole, was about to undergo a revolution. With Margaret Thatcher's general election victory, the fabric of Britain's economy was set to be torn down and radically remade. By the time she left Downing Street over a decade later, the Iron Lady had crushed the miners and sold off British Aerospace, Cable and Wireless, Jaguar, British Telecom, Britoil, British Gas, British Steel, British Petroleum, Rolls Royce and British Airways. Water itself would soon be privatised. Her successors would ensure the railways were also sold. However, even Thatcher herself saw limits to what companies should do. The KMS bodyguard contracts, which flourished under the Wilson/Callaghan Labour administration, caused the new Conservative Cabinet a surprising amount of consternation. Ministers could not understand why the Foreign Office was using private security guards, rather than serving British soldiers or policemen, to protect their diplomats.

The matter was discussed by a cabinet committee in January 1980, which carried out a short survey of bodyguard arrangements. They found that 'there is only one British firm which the Security Service [MI5] consider suitable – KMS Limited' – as a reliable source of commercial bodyguards. MI5's reasons for reaching this opinion of KMS are not given in the document, however it is an indication of just how close KMS was to the British establishment at the start of the 1980s. There were some logistical concerns, however, about the state's reliance on KMS. 'There must be a doubt in principle about the wisdom and propriety of relying on one rather small private firm for the security of government representatives overseas', a confidential briefing explained.

> Conflicts of commercial or other interests may affect the protection provided. One indiscreet or dishonest individual could do a lot of harm. So far KMS have provided excellent guards

at quite short notice (72 hours for men in Kampala) but this has depended on their other commitments, nor could they supply more than 15 or 20 men in all at present. This is just sufficient to cover our present requirements. Because many of their staff have an SAS background, use of the firm's services could stimulate added hostility from Irish terrorists in some cases. On the other hand, the Security Service advise[s] that the fact of a guard being a local citizen rather than a British ex-serviceman will not in itself inhibit such terrorists from making an attack.[37]

Other sources were considered, such as serving members of the SAS or the Royal Military Police, but both units were regarded as 'substantially under strength' and 'there could be difficulties in certain host countries in securing the necessary legal immunities', presumably should a guard kill a local person. (The meeting did not seem to consider whether any KMS staff were part-time soldiers, and what legal complications might arise from that. David Walker was still listed as a reserve member of the British army until the end of 1980.)[38] Both the Ministry of Defence and the police considered the role of guarding diplomats abroad to be beyond their normal duties, and creating a special unit from within the Foreign Office was something that the department's chief, Lord Carrington, considered 'absurd' given the 'very limited requirement'. Carrington was of the opinion that, 'Failing other solutions, it might be possible to continue to employ KMS Ltd bodyguards overseas. The main objection to this course of action lay in the past connection of that company with the provision of mercenaries in Angola and elsewhere' – suggesting that KMS had played at least some role in the debacle.

Thatcher summed up the brief discussion and said that the 'total number of bodyguards … involved was very small and the task not inappropriate to the Armed Forces. The Ministry of Defence ought therefore to be able to provide them.' Accordingly, the defence secretary was invited to 'reconsider whether and how the Ministry of Defence could meet the requirement to provide a small number of bodyguards for British Ambassadors abroad'. It was another bureaucratic fudge. Like the socialists of the 1970s, who threatened to ban KMS, the free market ideologues of the 1980s were lacklustre in their attempts to nationalise the nascent private security industry. The matter was pushed to one side and someone else was asked to look into it. After all, KMS occupied an important grey area, between public and private, that was rather convenient to preserve.

3

Teenage Rebellions

I never expected to share a taxi with a man named after a twentieth-century Soviet leader. And yet there we were together, Joseph Stalin and I, crammed into the back of a motorised rickshaw, charging through the mid-morning Colombo traffic. My guide, although an avowed communist, mercifully had none of the strongman character traits that one might associate with his namesake. The traffic was much more terrifying, like test driving a fairground ride. Finally, we pulled over on a quiet side street and darted down an alleyway. He had taken me to this unassuming house for a meeting with two old comrades, who were now in their late sixties. As we sat down for tea inside their study, the sound of Sinhalese and English soon filled the space. Their headquarters, a modest and humble affair, looked like it had not been renovated since the 1970s. The furniture was beyond retro, and the paperwork distinctly dusty. In better days, when these men were less jaded, political propaganda had sprung forth from these premises to enlighten the working class. As teenage revolutionaries, they had tried to bring down the government. But they had failed, and that failure was still haunting them.

In April 1971, this pair, and thousands of other teenagers and students their age, rose up in desperation at their appalling living standards. Their fury erupted in the south of Ceylon, that teardrop-shaped island at the foot of India. Independence from British rule had come almost a quarter century before, although the last troops only left in the mid-1960s. Their new rulers, despite promising a redistribution of wealth and better jobs, had done little apart from look after themselves. They spoke the language of socialism, while cutting the rice subsidy. So now the youth had armed themselves with swords and antique rifles. It was a motley collection of weapons, but enough to capture some isolated police stations where they could pillage the more formidable arsenals. Their movement was the People's Liberation Front, *Janatha Vimukthi Peramuna* in Sinhalese, or just the JVP.

These rebels were little older than children, and British diplomats described them as 'young unemployed university graduates from the rural areas'. Still, days after the uprising began, half a million bullets were on their way from a UK military stockpile in Singapore. Two thousand machine guns were also supplied. Britain's Ministry of Defence even pledged that 24 armoured cars would be 'despatched by first available ship', and sourced six Bell helicopters.[1] These efforts did not go unnoticed by the prime minister of Ceylon, Sirimavo Bandaranaike, the first female head of state in the post-colonial era. She told her British counterpart, the Conservative leader Edward Heath, of her 'very great personal appreciation of the action of your Government in responding promptly to our request for certain military supplies and making them available at such short notice'.[2] Later, the British envoy in Colombo commented that the key factor in crushing the revolt 'were the small arms, ammunition, armoured vehicles and other material supplied by Her Majesty's Government with quite remarkable speed and to decisive effect'.[3]

These weapon sales were announced in Parliament, but much more was happening behind the scenes.[4] In mid-April, with the rebellion in full swing, Whitehall feared the Ceylonese authorities were overwhelmed and close to collapse. Extreme measures were contemplated. Ceylon's army commander wanted help routing a rebel stronghold in Kegalle, an enchanting rural region. He claimed the rebels were hiding in 'isolated, unpopulated areas on jungled hill tops', and concluded: 'The only way to deal with them in his opinion and those of the other commanders was by Napalm.' The trouble was, Ceylon did not have any of this incendiary weapon, and hoped that Britain would supply some. The most senior Conservative politicians gave this request serious consideration. In the end, Foreign Secretary Alec Douglas-Home (a former prime minister), concluded that napalm could not be used effectively from a helicopter. He went on to explain that napalm 'is very heavy, needs skilled personnel to handle it and requires to be dropped with a high forward speed to achieve the necessary spread. It is also ineffective in jungle unless preceded by defoliation.' Britain's stockpiles were very low, and the foreign secretary did 'not wish to reveal the fact that we have any at all'.

However, Douglas-Home was still eager to assist Ceylon's military. 'An alternative to the use of Napalm would be the machine guns supplied with the Bell Helicopters', he advised the Ceylonese authorities. He also mooted that fragmentation bombs – prototype cluster munitions – could also be used. However, this idea was later scrapped for 'technical and presentational reasons', with ministers afraid of how the provision of such lethal weaponry might look to outsiders. If Ceylon's army continued to ask for alternatives to napalm, British military advisors were instructed to 'point out the usefulness of the machine guns which are already fitted to the helicopters'.[5]

These chilling diplomatic cables ran through my mind as I sat in that dusty Colombo office, talking to the survivors of this doomed uprising. In their teenage years, they could not possibly have foreseen the drastic response their rebellion would evoke thousands of miles away. At the UK National Archives, on the leafy upper reaches of the River Thames in Kew, there was a file called 'Material on the internal security of Ceylon – "Che Guevarist" insurgency'.[6] It was compiled by the Foreign Office's shadowy Information Research Department, a unit dedicated to countering communism. Its contents lay bare the coordinated effort made by the department to share all their anti-insurgency expertise, from Malaya to Greece, with Ceylon's rulers. In public however, British ministers played down their involvement in the conflict when questioned by conscientious politicians. A defence minister reassured Parliament: 'The Ceylon Government has not requested the provision of any British military personnel or advisers.'[7] In reality though, Britain's security service, MI5, had an officer, Mr Suckling, permanently stationed at the High Commission in Colombo. He had vast experience of undercover policing and the Ceylonese authorities acknowledged that they benefited considerably from his 'liaison on security matters'. To bolster Suckling's efforts, a British army colonel and a police-man were dispatched to Colombo for a week-long advisory trip because of doubts about Ceylon's ability to cope with the insurgency.[8]

A day before the additional British advisers landed in Ceylon, the High Commission warned that the local authorities seemed determined to 'destroy' the JVP with 'brutal and violent methods'.[9] Canadian diplomats were also concerned – not by the insur-gents, who they said were being careful to avoid atrocities against the rural population lest they alienate the 'peasants' – rather, the concern was that captured insurgents were being executed and that 'shocking excesses have occurred'. There were allegations of 'security forces shooting down innocent people and wantonly slaughtering insurgents. There are certainly very few prisoners.'[10] Ceylon's army commander justified these war crimes by saying the rebels were 'brutal people and we must be brutal too – I don't care if we have to destroy villages in the process'.[11] The military coordinating officer in Kegalle district, the epicentre of unrest, was a Sandhurst-trained lieutenant colonel, Cyril Ranatunge.[12] He told a Western journalist that he had 'already executed ten and would shortly be executing 12 more insurgents'. The corre-spondent who tried to file the story was told by the authorities to leave Ceylon within 48 hours.[13]

When the British duo of security advisers arrived in Ceylon, they walked into a police station where they saw 'three badly beaten-up insurgents who the police admitted had courage but little else. Apparently not many insurgents are being sent back to Colombo after capture.'[14] The scholar Fred Halliday noted: 'in later weeks hundreds of bodies of young men and women were seen floating down the Kelaniya river near Colombo, where they were collected and burnt by soldiers: many were found to have been shot in the back ... What is clear is that the police and armed forces launched an indiscriminate attack on the peasant population as a whole.'[15] Although Whitehall described reports of atrocities as 'disquieting', their officials had influential access to the highest levels of Ceylon's military and intelligence apparatus, and appeared to do little to stop the atrocities. The British colonel who visited the island said he was 'struck by the special relationship which existed between the British and the Ceylonese'. Many of the Ceylonese military leaders had originally been trained at officer academies in Britain and still held a deferential attitude towards the old colonial power. 'The links with the British Services are still very close', he effused. 'The officers I met were open, friendly, aware of their shortcomings, keen to improve and to benefit from British expertise and, rather like children, looked for support and encouragement. I believe there is merit in making conscious efforts to nurture this relationship.'[16]

* * *

Although many rebels were executed, and thousands are said to have died in the uprising, some 13,600 insurgents were kept alive in detention camps. This included the two gentlemen in front of me, who had welcomed me into their tumbledown study with such warmth. The mood was much darker now, as our attention turned to what happened after the uprising. These men said they were tortured in the detention camps, and 45 years later it was still too painful to describe exactly what they went through. As a sombre silence descended, I wondered why the British government of the day had gone to such lengths to crush their teenage dreams, and turn them into lifelong nightmares. Why, with so much access and influence over Ceylon's security chiefs, did Britain not urge moderation in their crackdown on the JVP? When challenged in the House of Commons about the appropriateness of arms sales to Ceylon that year, Foreign Secretary Alec Douglas-Home insisted: 'mediation is not required by the Ceylon Government, who are determined, if they can, to eradicate these extreme insurgents in their country'.[17] His statement alarmed the Labour MP Tam Dalyell, who argued, 'It is not good enough for the Ceylon Government to take the attitude that "mediation is not required". If we make available helicopters, I do not see why we should accept the brush-off. Further, what on earth is meant by your use of the word "eradicate"? We really should know a good deal more before supporting any move to "eradicate" anyone, even "extreme insurgents" with the use of British arms.'

Moderation was unnecessary because civil servants had made a simple calculation. 'From the point of view of both British commercial interests in Ceylon and our general politico-strategic interest', one wrote, 'the right course is to seek to preserve our influence by maintaining a generally helpful and sympathetic posture.' Foreign Secretary Alec Douglas-Home explained to the World Bank: 'it is not in the interests of the West that Mrs Bandaranaike's Government should be overthrown by left-wing extremists, we have therefore supplied military equipment and propose to go on doing so so far as we can.'[18]

Throughout the uprising, there was a constant Cold War concern that if Britain did not prop up Ceylon's unpopular leader, then the Soviet Union might swoop in to fill the void. 'Our policies are bound to depend a good deal upon our view of Ceylon's strategic importance in the Indian Ocean and our judgement of the implications for us of expansion in the influence of China or Russia', a British diplomat explained.[19] Just as Oman and Egypt occupied prime positions in world sea lanes, so did Ceylon, sitting at the centre of the Indian Ocean. Civil servants emphasised that, 'So far as the security of the Indian Ocean shipping lanes is concerned, our interest is that powers hostile to us should continue to be denied the use of bases in Ceylon'. And Britain was acutely aware of the value of having military bases on Ceylon. In the Second World War, after Japan captured

Singapore, Britain's Royal Navy moved its East Indies headquarters to Trincomalee Harbour, tucked away in the north-east corner of Ceylon. This sprawling maritime labyrinth is one of Asia's largest deep-water natural harbours, unique in the Indian Ocean for being accessible by all vessels in all weathers, making it the perfect place to park a fleet of warships. The legendary British admiral, Horatio Nelson, reportedly described Trincomalee as 'the finest harbour in the world' when he first set eyes upon it in 1775. The port is drenched with British naval relics. Amid the crumbling tombstones of long-forgotten English sailors, reindeer graze happily on the freshly cut grass. As alien to the island as the Royal Navy now seems, they were imported as pets by homesick seamen.

Although the Royal Navy had ceased using Trincomalee as a base in 1956, Prime Minister Edward Heath became personally concerned about the future of the harbour in the midst of the JVP uprising in 1971. He anxiously asked his intelligence chiefs what the consequences would be if Russia obtained the use of Trincomalee as a naval base. Their response was stark, stressing that access to Trincomalee would be of 'considerable advantage' to the Soviet Navy, who were otherwise reliant on distant Soviet ports for refuelling. The prime minister was warned that it would make it easier for the Russian fleet to occupy the centre of the Indian Ocean, 'through which many of the United Kingdom's and Western Europe's vital trade routes pass'.[20] The shortest route from cheap factories in China to Western markets saw thousands of container ships sail close to Ceylon, making it the fastest way to deliver their cargo. This briefing would have left no doubt in the mind of Heath, an experienced ocean yachtsman, that Ceylon was of great strategic naval significance to Britain. Other civil servants chimed in with this chorus of concern. Britain's future foreign policy towards Ceylon should ward off any hostile forces taking power on the island, both because of the risk that might pose to Indian Ocean shipping lanes, but also to avoid destabilising the much more populous Indian subcontinent.[21]

This was the geopolitical equation that meant the teenage rebellion would be crushed, whatever the cost. The men in front of me had paid a high price. School friends executed, their own bodies and minds disfigured for life. And yet in spite of all this suffering, in some senses they were actually quite glad that their uprising did not succeed. With a bitterness that is still raw, they regretted taking part at all. In their opinion, the JVP was not a genuine left-wing organisation. Although rhetorically it called on the workers of the world to unite, in reality it espoused a vicious hostility towards Ceylon's Tamil minority, around a quarter of the country. The JVP cadres were almost exclusively drawn from the island's Sinhalese majority, and their leaders saw the Tamils as 'alien invaders' from India. In fact, both ethnic groups can trace their arrival on the island back to roughly the same period, but this anti-Tamil racism is not uncommon among Sinhalese politicians. Such prejudice was one policy that the JVP shared in common with the 'bourgeois' rulers of Ceylon like Bandaranaike, who they had sought to overthrow. Bandaranaike realised the power of such a contradiction, and saw it as her salvation. It was evident that the economic situation in Ceylon was so dire the Sinhalese youths would rather risk death than toil any longer under these conditions. They needed an outlet for their anger, but the target did not have to be the Sinhalese ruling class. The JVP and Sinhalese working class resented the rich, but they could be made to hate the Tamils far more.

The potential for this kind of conflict, revolving around perceived racial differences rather than class, had long been apparent to Ceylon's politicians. In the campaign for independence from British rule, the Tamils in the north and the Sinhalese in the south of Ceylon were able to ally against the common colonial enemy. However, they had never lived together apart from under British rule. Prior to that invasion, the Tamils had their own kingdom in the north of the island, and the Sinhalese had several separate kingdoms in the south. If the island became independent as one political entity, then the Tamils would be a permanent minority in any island-wide elections – like the Scots are at Westminster. With this fear in mind, Tamil leaders such as the lawyer G. G. Ponnambalam sought assurances from Britain throughout the 1940s that there would be safeguards built into the independence deal: 50 seats in Parliament for Tamils and 50 for the Sinhalese. At a marathon session, he argued for ten hours that 'fifty fifty' representation was required, but his demands were roundly rejected by a constitutional committee chaired by Herwald Ramsbotham, the Right Honourable Lord Soulbury, an English Conservative peer. The Soulbury Commission, as it was called, naively asserted that there had been no proven acts of discrimination against the Tamils and there were unlikely to be any in the future.[22]

As it happened, one of the earliest acts of Ceylon's newly independent rulers in 1948, who were inevitably Sinhalese, was to deny citizenship to almost a million Indian Tamils who lived in Ceylon as tea plantation workers. They were some of the most exploited workers on the island, who toiled in conditions redolent of Victorian times. In denying them the right to live in Ceylon, the government had disenfranchised almost a million Tamil speakers with the stroke of a pen. Lord Soulbury's assertion that there would be no discrimination against the Tamils was rapidly unravelling. More blows would follow. In 1956, Ceylon's prime minister, Solomon Bandaranaike (who would later be succeeded by his wife), set about passing the 'Sinhala Only' bill, to eliminate Tamil as an official language on the island. This would severely disadvantage the Tamils, barring them from civil service jobs if they could not pass Sinhala language exams. In response, Samuel Chelvanayakam, the ascetic leader of a Tamil political party, mobilised 300 volunteers to stage a peaceful sit-down *Satyagraha* protest outside the House of Parliament in Colombo. As they sat like Gandhian disciples on Galle Face Green, a vast city-centre park next to the Indian Ocean, Sinhalese Buddhist monks began to beat them. The police looked on, and did nothing to intervene.[23] The horrific attack made the front page of the *Manchester Guardian*, under the headline 'Language riots in Ceylon'. Despite arousing international concern, the bill was passed into law, such was its domestic popularity among the Sinhalese majority.

This brazen assault on basic Tamil demands for equality, in broad daylight and in the heart of the capital, would not be an aberration. Instead it was fast becoming the norm. In May 1958, four days of rioting broke out. 'Sinhalese mobs went on the rampage, stopping trains and buses, dragging out Tamil passengers and butchering them', one Tamil author recorded. 'Houses were burnt with people inside ... Tamil women were raped and pregnant women slaughtered.'[24] The police once again did little to stop it, and eventually the army had to be deployed under a state of emergency. Hundreds of Tamils were killed in the rioting, and 10,000 made homeless. The situation continued to deteriorate in the 1960s. Sinhalese politicians competed for votes by running on anti-Tamil platforms, until even the Trotskyist LSSP, a major left-wing political party, also campaigned on this basis, alienating its Tamil supporters.[25]

After the JVP uprising was crushed in 1971, Sinhalese leaders accelerated the racial conflict, preferring it to a class war that directly threatened their own privileges. A new constitution was passed in 1972, renaming Ceylon as Sri Lanka, a Sinhalese name for the country. Buddhism, the religion of the Sinhalese majority, was given the 'foremost position' and the state was directed to 'protect and foster it'. No protections were made for Hinduism, Christianity or Islam, which were the most common religions among the Tamil-speaking minority. More policies were passed to discriminate against Tamil job prospects: to win places at university, they now needed to have better exam results than Sinhalese pupils. By the time the JVP uprising was crushed, Tamils were effectively second-class citizens. This placated many poor Sinhalese – they had just lost the class struggle, but at least they were winning the race war.

Tamil students, much like their predecessors in the anti-colonial Jaffna Youth Congress, would not accept such injustices and humiliation. In the early 1970s, they launched black flag demonstrations. Hundreds of Tamil students were arrested under emergency laws that remained in place from the JVP uprising. Yet they continued to protest, striking and boycotting all the schools and colleges in Jaffna, their northernmost city.[26] One young Tamil poet, Kasi Anandan, was detained for over 1,000 days, simply because he refused to swear an oath of allegiance to the new Sri Lankan constitution. Language was particularly precious to the Tamil people, being one of the oldest spoken on earth. Foreign linguists were increasingly recognising the importance of Tamil to human history and were keen to study it up close. While the Sri Lankan government was trying to stamp out this unique language, academics planned an international Tamil research conference in Jaffna. 'Hundreds of scholars from various parts of the world came to participate at the Jaffna conference. It was a historic and joyous event', a Tamil author wrote. On the final day, the seminar was open to the public and 'the assembled Tamils were listening in a state of rapture'. Then, 'hundreds of Sinhalese policemen threw tear-gas into the crowd and attacked the people'. Nine Tamils died, and hundreds were injured.[27] The government did nothing to condemn it.[28]

This onslaught almost inevitably made Tamils see coexistence with the Sinhalese as a dead end. Independence from British rule had only resulted in domination by the Sinhalese. Now Tamils wanted a state of their own. One of their main political leaders, Chelvanayakam, spelled this out when he won a by-election in 1975: 'Throughout the ages the Sinhalese and Tamils in the country lived as distinct sovereign people till they were brought under foreign domination', the veteran Tamil activist explained.

It should be remembered that the Tamils were in the vanguard of the struggle for independence in full confidence that they also will regain their freedom. We have for the last 25 years made every effort to secure our political rights on the basis of equality with the Sinhalese in a united Ceylon. It is a regrettable fact that successive Sinhalese governments have used the power that flows from independence to deny us our fundamental rights and reduce us to the position of a subject people.

His speech now reached its climax: 'I wish to announce to my people and to the country that I consider the verdict at this election as a mandate that the Tamil Eelam ["homeland"] nation should exercise the sovereignty already vested in the Tamil people and become free'. He made a solemn pledge to carry out this mandate.

It was a new dawn for Tamil politics, but the attacks continued. Sinhalese police turned their ire on Tamil Muslims, torching mosques in Puttalam and burning two men alive. All parts of the Tamil-speaking community in Sri Lanka were now under attack, from the plantation workers high up in the tea estates to the intelligentsia and youth in Jaffna to the Muslims scattered along the west coast. It was in this frenzied climate that Tamil leaders met at Vaddukoddai, an unremarkable seaside town in the north of Sri Lanka. Photographs from that day show Chelvanayakam standing on a small podium, wearing a light-coloured suit jacket, addressing the crowd in a field. Sitting on the grass near his feet, looking up and listening intently, are young boys, accompanied by their fathers. The crowd stretches into the trees in the distance, where people are standing patiently to see over the shoulders of those sitting in front of them. On this day, Chelvanayakam went further and was more explicit about independence than ever before. Drawing on the linguistic, religious and cultural differences between the Tamils and Sinhalese, he said Tamils were now 'a slave nation ruled by the new colonial masters, the Sinhalese'. He called for a 'free, sovereign, secular, socialist state of Tamil Eelam based on the right of self-determination inherent in every nation'. His demand for independence would ripple far and wide, and was beginning to be noticed by the old colonial power.

* * *

Hostage to Fortune

Iranian embassy, London, 1980

It was a Monday evening in May when the world watched members of Britain's most secretive army unit, the SAS, abseil down the front of the Iranian embassy in London. Inside, 20 hostages, many of them diplomats, cowered in terror as the six-day siege reached its climax. A desperate bid by militants from Iran's Arab minority to demand autonomy from the Persian clerics in Tehran was about to come crashing down. Explosions rippled through the building, doors were blown off their hinges and stun grenades were lobbed into rooms. Smoke billowed from windows where curtains had caught fire. Over the next eleven minutes, the SAS killed five gunmen, captured another and freed all the hostages bar one. It was the turning point in the regiment's history that propelled it out of the shadows and made it the world's most famous special forces unit.

For the 60 SAS men who took part, the mission remains an immense source of pride. Four decades later, these Iranian embassy veterans continue to carry a certain cachet that can almost guarantee them a lucrative after-dinner speaking invitation, or a spot on a TV studio couch as a counter-terrorism pundit. Behind this glamour, however, lies a less well-known story about what happened next to some of the soldiers at the heart of the siege.

As the rescue mission unfolded, a group of SAS men abseiled down the back of the building, aiming for a first-floor balcony. Among the climbers was Tom Morrell, a Fijian soldier who had joined the British army while his island nation was still under colonial rule.[29] He had years of experience in the SAS, but at the crucial moment of the descent his rope snagged, leaving Morrell dangling perilously twelve feet above the balcony. As he hung there, flames flickered up from the window beneath him. By the time Morrell's teammates on the roof managed to cut his rope, the Fijian's legs were blistering with burns.

Below the balcony, a reserve SAS team surged into the back of the building on the ground floor. Leading the charge was 30-year-old Rusty Firmin, a former marine, with shaggy red hair hidden under his hood. Behind him was one of the regiment's youngest ever recruits, 23-year-old trooper Robin Horsfall. Their team entered the Iranian embassy that day to find hostages already fleeing down the stairs in front of them. But the siege was not over. A gunman was trying to escape, disguised as a hostage and armed with a grenade. Firmin grabbed his arm and opened fire. Horsfall and others in the team also saw the threat and discharged their weapons. The hostage taker 'slumped to the floor with twenty-seven holes in him', Horsfall wrote in his memoirs. 'He didn't spasm or spurt blood everywhere. He simply crumpled up like a bundle of rags and died.'

*   *   *

Weeks after the Iranian embassy siege, in June 1980, a group of SAS soldiers arrived in Sri Lanka. Red-haired Rusty Firmin was among them, although this time there were no television cameras to broadcast the unit's endeavours. For the next four months they secretly trained Sri Lankan army commandos, selecting 60 members to form an elite anti-terrorist force. By the time they were done, Sri Lanka had two teams of commandos capable of mounting a counter-terrorism operation. The enemy was an armed group who, at that point, most people in the West had never heard of – the Liberation Tigers of Tamil Eelam, or LTTE. In time, they would become one of the world's most sophisticated guerrilla movements. But in 1980, when the SAS arrived, the Tigers had barely begun their campaign for independence from the majority Sinhalese state. And it would be another 20 years before they were officially designated as a terrorist group by the UK government.

When the SAS training was first proposed, Britain's defence attaché, Lieutenant Colonel Kenneth Reynolds, estimated that the Tigers only numbered about 20 'poorly trained young men'. They were so obscure that he called them by the wrong name, 'the Tiger Liberation Movement', and remarked that the 'terrorists are possibly little more than a gang of militarily untrained thugs'. The gang centred around a determined Tamil man from a fishing port at Valvettithurai (VVT), not far from the island's northernmost tip at Point Pedro. Velupillai Prabhakaran was the youngest of four children. He grew up in a relatively affluent Hindu family, with his father working as a civil servant. At school, he loved history, reading avidly about Indian independence leaders. For Prabhakaran, struggling against oppression was far from academic. He was just four when the anti-Tamil riots of 1958 ripped through the island. Growing up, he 'heard horrifying incidents of how our people had been mercilessly and brutally put to death by Sinhala racists'. He would later tell a journalist:

> Once I met a widowed mother, a friend of my family, who related to me her agonising personal experience of this racial holocaust. A Sinhala mob attacked her house in Colombo. The rioters set fire to the house and murdered her husband. She and her children escaped with severe burn injuries. When I heard such stories of cruelty, I felt a deep sense of sympathy and love for my people.[30]

As a schoolboy, he would have seen that whatever prosperity this family had was increasingly threatened by the discrimination against Tamils in jobs and education. He and his classmates joined the Tamil Students Organisation to protest against the pro-Sinhala policies, and their black flag demonstrations raced through Jaffna. But their grievances were met with arrest and torture.

Prabhakaran was indignant at what was happening around him. His elders, veteran politicians like Chelvanayakam, had spent decades struggling peacefully, within the electoral system and through the courts, but getting nowhere. They had been beaten and humiliated in plain sight. For Prabhakaran, now 17, these peaceful protest tactics were inadequate. In 1972, he formed the Tamil New Tigers with some close friends. A photo from the time shows him looking away from the camera, forlorn, burdened by the gravity of events. His black hair is trim, and a small moustache is beginning to grow. His shoulders are thick set, merging into his neck. He looks old beyond his years. Later, he posed in a flared black suit, hand on his hip, staring seriously at the camera. It was certainly the seventies, but he was in no mood to party.

As Chelvanayakam inched towards calling for independence, Prabhakaran's young crew leapt ahead of him. Using hit-and-run tactics, and bankrolled by heists, they were a fledgling vanguard waging a small-scale guerrilla war. Lieutenant Colonel Kenneth Reynolds summarised their activities as 'intimidation, murder and bank raiding'. In 1975, they assassinated the mayor of Jaffna, in revenge for the police attack on the Tamil language research conference. The hit made Prabhakaran a wanted man, but it showed there was another way to struggle. If the police killed Tamils, there would be severe consequences. The next year, even Tamil 'moderates' like Chelvanayakam finally looked beyond the status quo, with their open-air meeting at Vaddukoddai which had demanded self-determination. That same month, Prabhakaran also upgraded his strategy. He rebranded his organisation, forming the LTTE.

Although the Tamil militants were still few in number, British officials fretted that the Sri Lankan security forces were woefully ill-prepared to stop them. 'It would not require many more than perhaps 200–300 such men to give the Sri Lanka army an extremely "bloody nose"', warned Reynolds, who had spent almost three decades in the British army's Royal Artillery regiment before his posting in Colombo. 'I would go further by forecasting that I doubt the ability of the Sri Lankan army to contain such a terrorist threat.' Sri Lanka's generals, who ran an army that was mostly ceremonial, were not oblivious to the risk. From at least 1978, they had tried to establish an army commando unit at Diyatalawa – a breathtaking British colonial-era garrison town, tucked away in the central highlands – to combat terrorism and hijacking. But no soldiers would volunteer for the unit. Instead reluctant recruits were given a 35-day training course, which only 15 per cent passed.

The results were so disappointing that the whole 'company was disbanded, returned to unit, and a rethink in Army Headquarters took place', Reynolds commented. The exercise was repeated, with only marginally better results. The Sri Lankan army assured Reynolds that it now had 150 commandos, but the defence attaché commented sceptically: 'I have never actually seen the Commando do anything!' He was becoming concerned about the security of RAF flights that refuelled at Colombo airport on an almost daily basis, many en route to Hong Kong, which remained a British colony. He lamented that untrained terrorists could easily mount an operation at Colombo airport, jeopardising the RAF flights.

The defence attaché also had his eyes on the bigger picture – the Cold War. Reynolds noted that 'the option to use the Tamil Eelam [homeland] cause as a vehicle for fomenting unrest in Sri Lanka is wide open to Soviet agency activities'. He was suspicious of the Tigers' ideology, warning Whitehall that they had held 'indoctrination classes' in jungle villages and published a newspaper with 'overtly Marxist undertones'. By contrast to these shadowy subversives, Sri Lanka's newly elected Anglophile president, Junius Richard Jayewardene, appeared to have no problem with the former colonial power parking its air force on his country's soil. 'A year ago RAF Hercules aircraft transiting Colombo, being in camouflage paint, were required to park on the far side of the airport to avoid offending Sri Lankan sensitivities', the defence attaché recalled. After the election, 'These aircraft now line up proudly with civilian airliners on the international side of the airport with no murmur of objection.'

It was just the first sign of Jayewardene's fawning attitude towards London. Raised by an English nanny, he studied Latin at university and relished the trappings of British aristocrats. He had shown streaks of opportunism, supporting Britain's enemy Japan during the Second World War and converting from Christianity to Buddhism, a move guaranteed to win votes, but by the time he became president he was firmly wedded to the West. One British ambassador in Colombo later recalled to me how Jayewardene 'had this Romantic dream all along, that the best thing that could happen to you would be to be sat in a golden coach opposite the Queen and bowled down the Mall and treated to a state banquet at the end of it'.[31]

Reynolds picked up on this deferential mentality in Colombo and wanted Britain to take advantage of its historical connections with Sri Lanka's military. 'They are tied to our systems by virtue of their background and whilst they continue to receive training and advice from us it will be all that more difficult for the Soviets etc to penetrate or influence the Sri Lankan Army', he reasoned, recognising the importance of British military aid to prevent a rival power filling the void. By contrast, he scorned the Tamil demand for independence, even though it had a democratic mandate from the majority of the Tamil people. Chelvanayakam's Tamil United Liberation Front (TULF) had come second at the 1977 elections, making it the official opposition in Parliament. Their manifesto had fearlessly proclaimed: 'We alone shall rule over our land our forefathers ruled. Sinhalese imperialism shall quit our Homeland.' It went further and clearly spelled out that it sought 'the mandate of the Tamil nation to establish an independent, sovereign, secular, socialist state of Tamil Eelam'.

That election result terrified the Sinhalese, who felt threatened by Tamil political organisation and confidence. The next month, Sinhalese mobs spent a fortnight rampaging against Tamils, killing over a hundred. Reynolds had little sympathy, describing the Tamil campaign for self-determination as 'recklessly optimistic'. He regarded Jaffna, the Tamil's cultural capital, as the 'hot bed of Tamil fanaticism'. Possibly oblivious to pre-colonial times when Tamils governed a separate kingdom on the island, Britain's defence attaché asserted that an autonomous state would not be economically viable, something that was 'of little concern to the hot headed Tamil youth'. Reynolds blamed Tamil politicians for making 'inflammatory speeches' that were 'foolish and intemperate' and 'whipped up anti-Tamil feeling all over the country'. Without naming them, he clearly resented the aspirations of Prabhakaran and Chelvanayakam, whatever tactics they used. Even to speak of independence was insufferable.

Whitehall yearned for political stability on the island and offered SAS training to the Sri Lankan military during 1978. Initially, it was too expensive, but by the end of the year President Jayewardene himself was growing increasingly anxious about the deteriorating security situation. The LTTE had ambushed and killed a Sri Lankan policeman, Inspector Bastiampillai, who they blamed for torturing Tamils in detention. President Jayewardene soon enacted a law banning the LTTE by name, but the conflict was beginning to gather pace. At this juncture, he personally appealed to the UK ambassador for urgent advice from a British security expert. As it happened, Jim Callaghan's Labour government took the president's concerns incredibly seriously. The next month, MI5 sent an officer to visit Sri Lanka.

*   *   *

He was virtually retired. And in some ways, that made him more valuable. Now 67, John Percival Morton had accumulated a lifetime of experience in counter-insurgency. Better known as Jack, he really was an unrivalled expert in oppression. Born in India, he imbibed the racism of empire from a young age, believing that Indians were 'a sort of immature, backward and needy people whom it was the natural British function to govern and administer'.[32] Accordingly, he enrolled as a colonial policeman in India, where he spied on Gandhi's independence movement, then later headed up intelligence in Malaysia and became a key strategist in the war against the IRA. Now, in his twilight years, he would turn his hand to Sri Lanka. His brief for this trip was to 'report on the security organisation and recommend a number of changes and reorganisations at National Level to help combat terrorism'.[33]

As was usual for MI5, there was great secrecy around his visit. When I wrote to MI5 to ask for a copy of Morton's report, nearly 40 years later, they politely said, 'Your enquiry has been passed to the relevant department.'[34] That was the last I ever heard from them. The Foreign Office had a file called 'Sri Lanka: Security Assessment, 1978'. That file was destroyed by the department. Another record, 'Sri Lanka: Defence Visits from UK', was compiled in 1979, the year Morton visited. It too was destroyed.

We only know about Morton's visit to Sri Lanka because Britain's defence attaché, Lieutenant Colonel Reynolds, reported to both the Ministry of Defence and the Foreign Office, making two copies of all his letters. One set of Reynolds' copious notes survived the purge and can be found among RAF files tucked away in obscure parts of the National Archives repository. They cannot provide us with the full story, but they show that Morton recommended two spheres of military training that he considered to be vital to Sri Lanka's success: SAS training in anti-terrorism and anti-hijacking; and training military intelligence officers. This tactical combination, essentially a soldier-cum-spy, was a hallmark of Morton's long career in counter-insurgency. However, the Sri Lankans would still struggle to afford it. 'After a further period of procrastination and thumb sucking by the Sri Lankans and no progress, we virtually offered them the SAS training free of charge', Reynolds noted condescendingly. In July 1979, Sri Lanka officially requested SAS training and courses in the UK for three military intelligence officers. It was part of a concerted push by President Jayewardene to stamp out the Tamil uprising. That month, he also passed the Prevention of Terrorism Act. It was a draconian emergency law that allowed for 18 months' detention without trial. Confessions extracted while in detention would be admissible in court, giving police an even greater incentive to torture detainees.

Next, he dispatched a Sinhalese brigadier to Jaffna, a relative known as 'The Bull', under orders to eliminate the Tigers before the year was out. The army soon killed and mutilated two young Tamils, Inpam and Selvaratnam. Amnesty International said torture was used systematically in the offensive, 'including suspending people upside down by the toes while placing their head in a bag with suffocating fumes of burning chillies'. A Jaffna magistrate conducted an inquest into another dead Tamil youth, Indrarajah, and returned a verdict of homicide caused by the police. Jaffna was now under a state of emergency, and its youth were being rounded up and tortured. But this would not prevent Morton from visiting again. Besides, he thrived in such an oppressive environment. In November 1979, with the Conservative leader Margaret Thatcher sworn in as prime minister, Morton returned to Sri Lanka. This time, his focus was the police Special Branch and intelligence networks. 'Nothing but improvement can result from all this', the defence attaché noted excitedly, hoping that Morton would update and streamline the Special Branch.

Morton's racism might, ironically, have made it easier for him to advise Sri Lanka's police. The inspector general of police, Mr Seneviratne, was 'a Tamil hater', according to Britain's defence attaché, who was worried enough to conclude that it compromised 'his ability to make impartial judgement on the Tamil/ Sinhalese racial dispute'. Reynolds said the police chief's solutions involved 'fairly brutal retaliation' which was 'only further fermenting Tamil hatred'. Despite these seemingly serious reservations, Morton went on to advise a police force, headed by a 'Tamil hater', on how to reorganise its undercover policing capabilities to deal with the Tamil threat. Morton furnished the Sri Lankans with 'practical recommendations for the total reorganisation of the intelligence apparatus'. He supplied them with a 'Morton Report', which was at the 'heart of any discussion on Special Branch'. His report, which has never been made public, is said to have lamented 'the depressing picture of apparatus and morale in the security forces tackling the Tamil problem'.[35]

Gradually, his advice was implemented. In June 1980, an SAS team eventually visited Sri Lanka, fresh from the Iranian embassy raid, to advise the army commando unit there. Although the Foreign Office destroyed its file about 'UK military assistance to Sri Lanka' from 1980, which would have shed more light on this visit, one of the SAS team mentioned it in his memoirs. Rusty Firmin, a hero of the Iranian embassy raid, had been born in Carlisle 30 years previously and abandoned by his parents. He was adopted by a soldier, who sent him to join the army at 15, to stop him getting in trouble with the police. Firmin described the SAS trip to Sri Lanka in 1980 as 'an aspect of British "soft power". We generate a lot of good will by imparting useful skills and drills to foreign military units, we help spread British influence' – exactly as the defence attaché had envisaged.[36]

Firmin seemed to have little or no understanding of the specific political situation, other than the Tamil Tigers were a 'terrorist group which was seeking liberation for the sizeable Tamil minority within Sri Lanka'. While this might have given others pause for thought, Firmin ploughed on with the job. 'The training was all going to take place within Sri Lanka and we were given two months to get it done. The job was to teach them hostage rescue, and as part of their preparation they'd built a "killing house" and mocked up an aircraft fuselage at their Special Forces base on the jungle fringes outside Colombo.' The course included instruction on snipers, communications and using the Heckler and Koch MP5 submachine gun, as used in the Iranian embassy mission. The training was going well until one of the Sri Lankans forgot to clean his weapon properly and 'shot the guy sitting next to him in the head'. Firmin had to scramble to get an intravenous drip into the casualty while a helicopter arrived to evacuate him, but by the time he was seen in hospital it was too late. The man had died.

Firmin's memoirs are the only record of this incident. It is not mentioned in what remains of the official British archive, given that so many files from that year were shredded. 'Morale took a nose dive', we learn from Firmin. The careless commando was kicked off the team, but Firmin had to work hard to motivate the rest of the Sri Lankans to carry on with the training. Gradually, it came together. Soon, they were practising assaults on buildings, buses and aircraft; defusing booby traps and handling hostages after an incident. By the end, Firmin was pleased with the standard the Sri Lankan commandos reached. The team gave a demonstration to their top brass, who were impressed enough to put the SAS instructors up in a smart Colombo hotel, all expenses paid, 'where we could let our hair down and go and see the sights. All in all, it was a very satisfying experience.'

A year after the SAS training, their Sri Lankan protégés staged a 'well mounted covert operation', ambushing Tamil militants, killing two and capturing one. Lieutenant Colonel Cox, who took over as Britain's defence attaché in Colombo, noted with satisfaction that the commandos had 'gained revenge' for earlier attacks, in which two soldiers out shopping were killed and another had died while trying to stop Tigers robbing a bank. The Tamils saw it very differently. That summer, the police and army once again went on the rampage in Jaffna, on the eve of an election. Over a hundred shops were ransacked, and market squares looked as if they had been shelled. Several people were killed by the security forces, and the headquarters of Chelvanayakam's old party, the TULF, were burned down in an extraordinary attack on the country's official opposition. For the Tamil people though, there was one building whose destruction was unforgivable. The public library in Jaffna held over 90,000 volumes of text, some so old that they were written on palm leaf. It was razed to the ground by arson. Attacks on the Tamils continued, leaving one observer to call it a 'cruel summer of murder, arson, pillage and plunder'. By the end of the year, with his reputation in tatters, President Jayewardene hired a London-based PR firm to burnish his image – the company's previous clients included the Shah of Iran and apartheid South Africa.

Jayewardene was becoming increasingly concerned about Tamil activists both inside and outside Sri Lanka. Many were fleeing to the UK to seek asylum and lobby for their cause. By December 1981, MI5 was keeping a 'close eye' on Tamils in London.[37] One group under surveillance was the Tamil Coordinating Committee (TCC), 'a small group of Tamil residents in London who produce skilful propaganda but who, according to the Security Service [MI5], have little capacity to mount demonstrations'.[38] The British authorities were familiar with the key Tamil activists in London, including the son of the TULF leader, and a feud he was having with the TCC. They noted that although the TCC had 'Communist connections ... there is no evidence of Communist funding of the TCC. (The most likely source is wealthy Tamil businessmen.)'[39] Amid this deep analysis of Tamil diaspora politics, the Foreign Office reminded civil servants of 'the importance to us of Sri Lanka's stability and continued pro-western alignment'.[40] Britain would continue to thwart the Tamil independence movement through increasingly elaborate channels, with MI5 and SAS stalwarts such as Morton and Firmin slowly giving way to their former colleagues, men like Horsfall and Morrell, who would technically no longer be working directly for Whitehall.

4

The Upside Down Jeep

As British spies and soldiers increasingly focused on crushing the Tamil and Irish independence struggles, the Dhufari liberation movement earned a brief respite. The last official SAS team left Oman in May 1978, leaving 55 known revolutionaries still at large. Hiding deep in the mountains of Dhufar, the resistance was down but not out. A month after the SAS departed, they attacked civilian and military targets in eastern Dhufar, killing seven and wounding eight more. Their summer offensive 'caused a good deal of unrest among the expat community', commented Britain's defence attaché in Muscat. Hawkishly pro-Sultan, Colonel Knocker monitored rebel activity and remarked on 'two spectacular and effective actions by the remaining enemy'.[1]

Their main attack was on a beach party at Taqa, between Salalah and Mirbat. Five contractors for the British defence company Airwork were killed as they enjoyed their holiday. The beach trip had become a regular festival, allowing the resistance to track their movements and plan the assault. The other two casualties that year were soldiers in Oman's artillery, who died in an ambush. Colonel Knocker commented that 'the enemy can and will attack soft targets, and gain considerable kudos from such actions'. The revolutionaries trumpeted their victories on Radio Aden, over the border in Yemen.

The Sultan tried to extract his revenge in the autumn, with a large-scale operation, but failed to find any rebels.[2] British officials complained that the enemy looked the same as the local population, among whom they clearly still had some support. Even the defections were yielding little in the way of accurate intelligence. The 'hearts and minds' strategy, to develop Dhufar and dangle a carrot among the inhabitants, was mainly of benefit to the military. Few civilians had cars, but the new roads were particularly helpful for troops to resupply remote outposts by land, when it was too misty during the monsoon season to fly in stockpiles. The relative peace in Dhufar did see some refugees and ex-fighters trickle across the border with Yemen, sometimes two or three a week, and occasionally up to 150 at once. Often they were women, children and elderly men, the relatives of revolutionaries, and they were becoming a burden on the Yemeni authorities.

Britain still played a major role in Oman, even if the SAS no longer had a permanent presence. 'British mercenaries tend somehow to hang on and indeed dominate the Omani scene, which seems to be more than acceptable to the nice Omanis', the deputy head of Oman's Defence Ministry noted.[3] Some of the most decorated SAS Dhufar veterans now served the Sultan under the auspices of KMS Ltd, in what was the company's most valuable contract. One British official later commented: 'it is a reasonable supposition that this contract has been a major factor in their [KMS] financial viability'. The deal was done as early as 1976 and saw the company set up the Sultan's Special Forces at Zeek in Dhufar. This was an elite unit that was commanded by KMS personnel, who also provided its key training staff and provided all logistical support, 'from pencils to machine guns, Land Rovers and boats'.[4]

Several well-connected war veterans had been involved in founding this mysterious company: two majors, David Walker and Andrew Nightingale, the old brigadier, Mike Wingate Gray, and that pioneering mercenary Colonel Jim Johnson. The renegade colonel was the product of empire. One of his forebears had served as a soldier in the East India Company, and the young Jim was raised in Ceylon where his parents owned tea plantations. The family moved to Britain where the father worked on the Enigma project during the Second World War, and the son attended the prestigious Westminster School alongside the Labour left-winger Tony Benn, but the pair went on to follow very different paths in life. Although both men enlisted to fight for their country in the Second World War – Benn flew in the RAF and Johnson fought for the Welsh Guards – that is where any similarities end.

After seeing some action towards the end of the war, and dogged by injuries and illness, Johnson seemed to have lost interest in the military, and by 1947 he had moved into the world of finance. He became a dapper insurance broker working for Lloyd's. At the same time, he enjoyed intimate connections with powerful commanders such as David Stirling, who had founded the SAS during the war. For the next 15 years, Johnson led a double life, a city slicker who served simultaneously as a reserve officer in the SAS. Little is known about what he did for this elite regiment, but he was clearly one of its best officers, rising to command the reserve unit by 1960. Rare footage from that year shows Johnson leading a parade of SAS reservists as they march through Piccadilly to lay a wreath at a war memorial. The stocky colonel then swaggers around the parade square, smugly inspecting the dozens of men under his command. He remained in charge of this reserve special forces unit for another two years. Thereafter, he supposedly retired altogether from the army, but was not forgotten by his former colleagues. When the royalist regime in Yemen was toppled by Nasser-backed revolution-aries, Stirling met the head of the SAS in a London club and, with the connivance of several Conservative ministers such as Amery, decided that Johnson should be mobilised in an unoffi-cial capacity to help restore Yemen's royal family to power. He is reported to have accepted the offer by saying: 'I've nothing par-ticular to do in the next few days. I might have a go.'[5]

From 1963 to 1967, Johnson led a rag-tag band of British and French mercenaries on sabotage operations against Nasser's allies in Yemen, and supported the royalist militias to push back against republicanism. Some of Johnson's men were quietly released from the SAS to join his campaign, and MI6 intermittently lent their substantial support. By 1964, the covert war in Yemen was becoming so significant that MI6 reviewed its policy on 'deniable operations' in light of challenges posed by 'de-colonisation and newly emergent nationalism'. At a top secret meeting, Whitehall's Joint Action Committee reasoned: 'An operation is deniable if, in spite of the probability that HMG connived in its execution and in spite of some tenuous or arguable evidence that HMG was officially involved, HMG considers it politically feasible to deny complicity in public statements, e.g. in the House of Commons or the United Nations'. MI6 noted that it was 'equipped to carry out small-scale deniable operations on its own or to make a significant specialist contribution to larger scale deniable joint operations,' and that cooperation between MI6 and the military in this field was 'growing and is likely to grow still more'.[6] When the mission in Yemen was becoming too tiresome for Johnson to keep up his duties at Lloyd's, the MI6 controller for the Middle East and Africa asked the company's chairman to grant him leave, a telling example of Johnson's connections to some of the most secretive parts of the British state.[7] After this covert war was over, Johnson spent five years as Queen Elizabeth's aide-de-camp, a prestigious ceremonial role with the royal family that showed his unconventional actions in Yemen had only brought him closer to the British establishment. A year after leaving the Queen's side, Johnson decided to formalise his chosen line of business under the banner of KMS. He was quickly able to impress his new colleagues with lucrative contracts to train Yemen's presidential bodyguard and assist Saudi royals, who had supported his earlier campaign across the border in Yemen.

Johnson also brought to the table formidable domestic political connections in Westminster, which would endure into the 1970s and 1980s, as a curious incident involving a London museum demonstrates. In 1977, the National Army Museum in Chelsea was seeking to expand and needed donations to fund it. A Conservative MP, Sir Anthony Royle, was asked to contact the Sultan as a potential donor. Sir Anthony had been a foreign minister in the early 1970s, at the height of the Dhufar war, and would go on to be vice-chair of the Conservative Party under Margaret Thatcher. He took the museum's request very seriously, and insisted that no one else contact the Sultan until he had done so.

Within six weeks, the approach was bearing fruit. A memo in the museum's archive notes: 'Sir Anthony's emissary, Colonel Jim Johnson has just returned from Oman. He approached the Sultan with a request for a contribution to the museum's Building Appeal. The Sultan said he would give £150,000.' This sum would now be worth around £650,000, and in any event covered a quarter of the museum's building costs. It was a crucial gift. The memo went on to explain cryptically that 'Colonel Johnson serves the Sultan in a variety of capacities', and it also revealed that the colonel had been Royle's best man. In other words, this KMS founder was intimately connected to a rising star in the Conservative Party and trusted by several Gulf rulers.

The gravitas of Colonel Johnson's political connections became apparent at a meeting held shortly before Christmas 1977 with the National Army Museum's directors. Colonel Johnson was flanked by his business partner, David Walker, who the museum believed was his 'assistant' and served in Johnson's 'SAS squadron in Oman', a confused reference to the KMS team's work there. This curious gathering of men took place at an exclusive west London terrace, which was the office of none other than Brigadier Tim Landon. The White Sultan himself attended the meeting, where he reported that Qaboos would be 'very happy to contribute £150,000' in appreciation of the links between the British army and Oman. This episode shows that KMS did not only command the Sultan's Special Forces. The company worked in tandem with the White Sultan, who relied on them to disperse Qaboos' money wisely on cultural institutions. They were expected to be renaissance mercenaries, experts on both war and art. When the construction project was approaching completion in 1979, it was Colonel Johnson who conducted a site inspection. Although the museum still acknowledges the sponsorship it received from Sultan Qaboos, the role played by British mercenaries in arranging this deal, and their close association with Landon, has remained completely hidden for 40 years.

Below the KMS elite of Landon, Johnson and Walker, there was a tier of SAS veterans who focused on the day-to-day grunt work of training the Sultan's Special Forces. Among these KMS instructors was an Oxford graduate in politics, philosophy and economics (PPE), the favourite course of Britain's ruling class – although Major Shaun Brogan was something of a maverick. Born into a military family, he hated blood sports and loved music and poetry. He rebelled at Sandhurst by driving his girlfriend's car bearing the banner 'Make Love Not War'. After incurring the commandant's wrath, he reluctantly conformed, passing out as an officer and later earning entry into the SAS.

Despite his youthful hippy sympathies, he clearly had no love for the real revolutionaries in Dhufar, where he would zealously make war. Arriving in mid-1971, he found that a few hundred rebels had already defected, lured by the new regime of Sultan Qaboos. Brogan's task was to recruit and train these defectors to go back up into the hills and fight against their former comrades. Once the training was complete, he led them into battle, flying in by helicopter. They came under heavy fire and had to withdraw across a swollen river at night. Brogan's inexperienced local troops were terrified and sought cover, but through what military records call his 'determination and personality' he persuaded them to advance. British pilots encouraged them further by bombing the rebels below, and after seven hours of continuous fighting Brogan's team had killed the local rebel commander. 'Couriers and sympathisers were uncovered and punished', his army citation noted ominously. The mission managed to clear 350 square miles of rebel-held territory, winning him the prestigious Military Cross. Later that year, Brogan was wounded in action and took two months to recover, before heading straight back into the conflict.

Brogan, like David Walker who served in Dhufar shortly after him, was intellectually capable. He punctuated his military career with studies at Oxford, where he met his wife, a young doctor. On his return to Oman with the SAS for a second tour, she joined him as a flying doctor, effectively an extension of the hearts and minds mission. Brogan then moved seamlessly from the SAS to working as a mercenary in KMS, staying in Oman to train the Sultan's Special Forces for two years in the late-1970s. The couple eventually moved back to the UK and became heavily involved in the National Health Service, before Brogan passed away in 2017, with his memoirs drafted but unpublished. His widow is staunchly proud of his achievements in Oman, writing in a tribute that the Dhufari defectors 'did respect him as did all the tribes men especially in the Western area where we worked together to win the peace'. Brogan's book, if it is ever published, will probably add to the hagiography of British military missions in Dhufar, but these veteran memoirs rarely make space for the perspective of those that they were fighting against.

For instance, while Brogan was training the Sultan's Special Forces in 1979, the Dhufari revolutionaries were struggling on, nearly a decade after the supposedly benign Qaboos had come to power. The resistance, clearly still aggrieved, managed to stage 14 attacks that year. A rate of at least one a month was an impressive feat for a movement that had been so extensively penetrated, split, bribed and bombed. They continued to pose a lethal threat, killing six of the Sultan's soldiers and wounding nine, and managed to keep hidden a formidable arsenal of mines, Kalashnikovs and rocket-propelled grenades. The rebels used two pounds of plastic explosives to blow a hole in the dish of Oman's Earth Satellite Centre at the start of 1979. Roads were still being mined, blowing up an artillery vehicle and maiming a commander, but also killing a local Dhufari woman.

The revolutionaries even maintained a foothold on Jebel Aram, where KMS manager David Walker had fought with such distinction as an SAS officer many years earlier. There was another battle there in 1979, which was particularly poignant. The revolutionary leadership were said to be meeting to plan operations to mark the anniversary of the Dhufar war. Armed with AK-47s, they were caught in a firefight with the Sultan's Frontier Force.[8] Three rebels, including two of its leaders from the central and eastern areas of Dhufar, were killed in combat. A fourth rebel was trapped in a cave, from where he managed to keep up 'heavy and accurate fire at the surrounding troops'. The commander of the Sultan's patrol, a New Zealand mercenary, Major Donald Nairn, 27, was worried the rebel might escape into the night, and decided to 'smoke him out'. This was to prove fatal for Nairn, who was shot and killed by the rebel as he advanced towards the cave. It was a pyrrhic victory for the resistance though, as their gunman later surrendered to the Sultan.[9]

* * *

The Dhufari revolutionaries could not be allowed to stage a successful fightback. The revolution in Iran and the fall of the Shah in 1979 had increased international attention on the region, especially on Oman's northern tip at Musandam, a barren peninsula. 'This area, with its command of the oil-important Straits of Hormuz, is what the Dhufar War was all about', Britain's ambassador in Muscat advised the foreign secretary.[10] 'Its significance has increased, partly as a result of the abandonment by Iran of her former role as gendarme of the Gulf'. The Sultan of Oman would now have to assume an exclusive role as the West's custodian of this oil supply artery. Landon lamented that 'events in Iran are, indeed, a sad example of the world-wide decline in the non-Communist community'.[11] As the world entered the 1980s, the Dhufaris only had 45 fighters left, having lost ten in two years. The Sultan was determined to wipe out the rest of them within months, and his army commander set a deadline for the start of April. He unleashed Operation Southern Comfort, which aimed to 'locate and destroy the remaining enemy in eastern Dhofar [*sic*]'.

The resistance, sensing a massive military build-up, preferred to melt away into the middle distance and lie low. It was only after the offensive, as the sun set on 28 March 1980 and the deadline loomed, that there was finally a firefight – killing one rebel, wounding another and resulting in the surrender of a resistance leader. The military saw this as a 'significant breakthrough which is due to the continuous pressure which is being applied to the enemy'. The latest defector was Salem Ahmed Mohammed Al Sail Al Ghassani, a member of the revolution's central committee. He appeared on national TV and pronounced that the popular liberation 'front' would become a 'party'. Although tamed, his politics still posed a rhetorical threat to Qaboos. He called for democratic liberties, the release of political prisoners and the expulsion of all foreign forces from Oman. His anti-imperialist vision included a non-aligned foreign policy, commitment to Arab nationalism and solidarity with the Palestinians.[12] Despite his defection, by the end of 1980 the resistance still mustered 40 to 50 cadres somewhere in the eastern hills of Dhufar, and the deadline for their destruction was now dismissed as an April Fools trick.

The role of KMS and the Sultan's Special Forces in this relentless counter-insurgency operation is hard to fathom based on the British archives. It seems that Landon had so much control over these elite warriors that British bureaucrats in Whitehall were assiduously kept out of the loop. The Sultan's Special Forces were a rule unto themselves, receiving lavish amounts of money and representing a significant power block within the military.[13] Few insiders are willing to speak about what the company did in Oman. A widow of one KMS man first responded warmly when I approached her, and readily offered to meet me the next day. While we were finalising arrangements over email, she asked what exactly I was writing about. As soon as I said KMS, the emails went cold.

Later the next day she returned my calls, addressing me in a cut glass Home Counties accent. 'I Googled you and see you write a lot of articles', she said anxiously. 'You're an investigative journalist is that right?' The phone call did not go well. 'It's just useful to have a bit of background because you've already dis-covered the sort of company that it is – You're not working for them?' She inquired, always calling them 'the company'. After I denied having any connection with the company, she said: 'It definitely requires a lot of confidentiality and anonymity if I do decide to talk about things', adding that her late husband would 'have been much more likely to either give you a picture, or not give you a picture, of what went on'. More questions followed about my angle for the book, and what motivated me to write about the company. 'What I'd like to do is think about it … leave it with me for a couple of days and I will come back to you', she assured me in the end. Weeks passed with no response, until I eventually called her back, expecting another grilling. This time, the line was comically crackled, and she apparently no longer had much recollection of KMS, so there was no point me asking her about it. She rang off, as I marvelled at the way people could suddenly 'forget' what they knew about the company.

* * *

2 May 1981. A Range Rover raced along the newly built road in Dhufar, trying to catch a flight that would leave soon from Thumrait, a dusty oil town with a military runway scratched out of the desert. In the passenger seat was a director of KMS, Major Andrew Menary Nightingale.[14] These days he liked to call himself a security consultant. He was looking forward to returning home to Putney Heath, one of the greenest parts of south-west London, within walking distance of Richmond Park, Wimbledon Common and the Thames.[15] He had spent a decade

as an officer in the British army's intelligence corps, winning an MBE for his exploits in Northern Ireland during 1977 before he retired, joined the reserves and helping to found KMS and its sister company, Saladin Security. In his new corporate role, he had been trusted by the Foreign Office to protect Lord Carver, Britain's resident commissioner in Rhodesia, in 1979, amid deadlocked negotiations over independence. The firm had also sent him out to Dhufar to command the Sultan's Special Forces, 'a battalion sized unit predominately made up of Dhofari [sic] tribesmen'.[16] The unit was modelled on the SAS but was also 'trained to carry out additional overt and covert operations'. It was packed with approximately 20 ex-special forces officers from Britain and other countries serving in the unit. But now it was time for Nightingale to hand over control of this elite squad to the gentleman chauffeuring him to the airport, another retired major, 37-year-old Julian Antony Ball.[17]

The new commander of the Sultan's most trusted unit was an equally decorated war hero. Slim, with a thin face, he exuded nervous hyperactivity as he constantly puffed his trademark Capstan Plain cigarettes.[18] Ball came from a wealthy family and spoke with an upper-class accent. His parents had sent him to the army's sixth-form college at Welbeck, but financial difficulties caused Ball to drop out before taking his exams. His wealthy grandfather had passed away, and left his inheritance to a friend rather than the family. Ball enlisted in the army as a private in the Parachute Regiment, and began to climb through the ranks. In these working-class circles, his first name sounded too posh, so he went by Tony.[19] He made his way into the SAS, and by 27 became an officer in a Scottish infantry regiment, winning a Military Cross for his action on the streets of Belfast in 1970. Preferring the culture in the special forces, he often went on covert missions in Northern Ireland, even before the SAS was officially deployed. He worked in the SAS counter-revolutionary warfare wing, training soldiers to work undercover in Ulster with 14 Intelligence Company, a covert reconnaissance unit.

An admiring colleague said Ball had 'an affinity with this sort of work'.

Others have been more critical of Ball's service during the Troubles, claiming that he tutored and oversaw a notorious captain, Robert Nairac, who is alleged to have taken part in sectarian murders along the border. Nairac's life and later killing by the IRA is shrouded in myth and legend, making it hard to discern what incidents he was really involved in. One of the more consistent claims is that while under Ball's command, Nairac ventured across the border into the Republic of Ireland and took part in the fatal shooting of an IRA man, John Francis Green, during a ceasefire in 1975.

Whatever the truth, Ball remained part of a trusted special forces elite. When Prime Minister Harold Wilson eventually deployed the SAS to Northern Ireland on an official basis in 1976, Ball was one of the first on the ground, before comrades like Brian Baty or Malcolm Rees could join him. A colleague claimed that some soldiers 'were petrified that we would start shooting everything and everybody'. Ball was constantly on edge, living in a small windowless bunker at the Beesbrook base, and kept his room-mate awake with 'regular nightmares'. Like many in the SAS, Ball was 'not afraid of authority and would frequently bend the rules if he thought it necessary'. However, his friend Nairac, who was not in the SAS but worked closely with them, eventually went a step too far. He had a penchant for visiting republican bars in South Armagh, ostensibly to gather intelligence. He rarely told superiors what he was doing, and one night in 1977 he was abducted and murdered by the IRA. His body was never found and has become one the most contentious killings among British army veterans – many claim the IRA fed his body into a meat mincer, a theory ruled out by the head of the Independent Commission for the Location of Victims' Remains. Understandably though, Nairac's death and disappearance affected Ball deeply. He missed his maverick comrade and was vexed at how his intelligence operations were managed. Ball reportedly turned his home into a shrine to Nairac.[20] By 1980, Ball had had enough of the British army. He had earned an MBE, but felt his lowly background as a private would always be 'held against him'. This sense of resentment led to him driving this Range Rover through Dhufar, where the subtleties of the English class system would be lost on the local troops he now commanded. He fitted in perfectly at KMS, who had found that members of 'Army Intelligence, and its associated operations in Northern Ireland' were a 'particularly good source of recruits'.[21] Indeed, the job as a KMS mercenary came with a promotion to lieutenant colonel in the Sultan's army.

* * *

There were no survivors. No inquest. And no witnesses have ever come forward. The Range Rover was upside down, not far from the airfield at Thumrait. Both men in the car were dead. Perhaps Ball had taken the corner too fast, rushing so Nightingale could catch his flight home to Putney? Although neither man was still in the SAS, their old regiment claimed the bodies. The funeral at Hereford was a military affair, with plenty of ceremony. Snipers sat snugly in St Martin's church tower, on the look out for IRA assassins. Senior officers and notables attended the event. Saladin Security transferred Nightingale's share in the company to his executor, Margaret. The Ball family would have preferred a more intimate send-off, friends and family only, followed by cremation and scattering of ashes. Instead the SAS buried him in their plot, claiming him as one of their own for eternity, his name and old regiment chiselled into the headstone. He had died serving the Sultan as a mercenary, but at the end of the day the SAS could not bear to let him go. The connection between Keenie Meenie and its member's old regiment was seemingly unbreakable.

* * *

The tragic and mysterious death of two top KMS commanders in Dhufar did not dampen the company's prospects in Oman. Their benefactor, the 'White Sultan' Tim Landon, was still very much at Qaboos' side. Six months after the car crash, he met with the new British ambassador. The Foreign Office was anxious to ensure the pair got off to a good start. There is no mention in the meeting's minutes of the road accident, and searching through the archives it is as if it never happened. Indeed, Landon was relaxed at the meeting. He was increasingly pursuing his own business interests, as well as working for the Sultan, who often sent him on missions to the United States. 'Landon said the Sultan had grown greatly in stature during the last two years as a Ruler. He was now much more confident and enjoyed giving a lead', the Foreign Office noted, rather suggesting that for the first decade of Qaboos' reign, Landon had been firmly in charge.[22] Qaboos was an ardent Anglophile, and was looking forward to a state visit to the UK at the end of the year to see the Queen. The Sultan would welcome going to an opera, 'but *not* a modern opera by, for example, Benjamin Britten', the diplomat said. 'He would also enjoy a military parade or occasion such as [the] Beating Retreat. Brigadier Landon mentioned that at some later date the Sultan would be greatly pleased by an invitation to take the passing out parade at Sandhurst. He was envious of [Jordan's] King Hussein's having done this. Such an invitation would be appropriate since he had done a good deal for Sandhurst and Sandhurst had done much for him.'[23]

The situation in Dhufar was calmer now. There would be no contact with the enemy for the next two years. By the end of 1983, the revolutionaries only just 'remain an enemy and a small core of fighters numbering around 20 still roam in the Eastern Dhofar [*sic*] mountains'. And yet, despite all the Sultan's powerful friends, it seemed impossible to kill an idea. The revolutionaries 'continue to be trained in Eastern Bloc countries and radical states such as Libya and Syria', the defence attaché in Muscat warned. 'Only some 500 of an estimated 2,000 of its supporters returned to Oman in a period of grace offered by the Sultanate (up to 30 April 1983 but extended indefinitely). Since the extensions, there has been a mere trickle.'[24]

* * *

Although Landon continued to prosper in Oman at the expense of the revolutionaries, the company he was associated with was beginning to flounder. Aside from training the Sultan's Special Forces, and a smattering of bodyguard contracts with other Arab royals, KMS was struggling to find much work. Thatcher's Cabinet had proved surprisingly proactive at bringing the embassy bodyguards in-house, and by the start of 1982 KMS had lost almost all of its contracts with Britain's Foreign Office. The European Commission had emerged as an unlikely new client, hiring a five-man armed KMS unit to guard its delegation in Uganda for £117,740 a year (now worth more than £408,000).[25] However, the company was struggling to satisfy its customer, and one European delegate complained about the high turnover of KMS staff who were all 'virtually "wedded" to the SAS', noting that their 'highly specialized military training ... does not seem to be ideal for the civilian employment such as they now have'. He complained that the company's senior staff were 'useless' at enforcing discipline with the result that the KMS men had been involved in 'several major car accidents, drunkenness, etc.' More seriously, their 'professional competence' itself was being called into question after KMS staff 'failed' to deal effectively with two vehicle hijackings.

With its contracts dwindling or coming in for criticism, some senior figures in the company appeared to be putting KMS on the back burner. After a brief business trip to Uganda in April 1982 to meet KMS colleagues 'bedevilled' by problems on the European Commission contract, David Walker seemed to be contemplating a new and altogether more respectable career. On 6 May 1982, he was elected as a Conservative councillor to represent the ward of Esher, a rural Surrey town in the borough of Elmbridge. He was one of three councillors in his ward, where the electorate numbered 4,310 people.[26] He could not have picked a sleepier place to become a politician, far away from the bloody strife of Dhufar or any of the other conflict zones where he had made his money. Indeed, his whole domestic lifestyle was totally at odds with his endeavours abroad. He lived in a luxurious detached house called Chestnuts, at 3 Sandown Avenue, a gated private road with some of the most expensive properties in Surrey and a short walk from Esher Cricket Club, an artificial ski slope, a golf club and a racecourse. It was the quintessential Surrey commuter town for people who worked in finance in the City of London.

A fortnight after the vote, Walker strode into Walton-on-Thames town hall, and eagerly volunteered to sit on two committees: Town Planning, which would suit his engineering degree; and Recreations and Amenities, which might have been more of a challenge for this security expert. Within days, he was assiduously wading through pages and pages of planning permission proposals with his fellow councillors. They examined a plan to install a new bar at a hotel in Weybridge, a fire escape for a restaurant on Cobham high street and a refurbished kitchen for a pub in Hersham.[27]

Meanwhile, thousands of miles away, Britain was once again at war. This time the battle was deep in the South Atlantic, over the fate of 2,000 British citizens, less than half of the electorate of Walker's ward. A British task force was furiously fighting Argentina over the Malvinas/Falkland Islands. Walker's election victory came barely a few days after the *General Belgrano* was sunk, and while British special forces were covertly infiltrating the islands. One of Walker's first assignments with KMS had been to supply bodyguards in Buenos Aires to protect British diplomats, a successful service Whitehall had not forgotten. This time, the Foreign Office was concerned about the safety of its diplomats in neighbouring Uruguay, who were being targeted by the Argentine government. Four armed KMS guards were quickly dispatched to Montevideo to defend the embassy from such an attack.[28] The team remained there until August 1982, under great secrecy. At a council meeting after the islands were reconquered, Walker and his fellow councillors expressed 'sympathy with all who have been bereaved', as some local men had died in the battle, and 'recorded their admiration for the way in which the country's armed forces and the merchant navy have conducted themselves in the successful restoration of the Falkland Islands to British sovereignty'.[29] However, his own company's role in the conflict remained hidden and the council quickly moved on to other matters, congratulating Prince Charles and Princess Diana on the birth of their first son. It was all terribly civilised.

Walker approached his new life as a small-time politician with gusto. In the first 18 months, he attended 30 meetings, missing only four. He scrutinised the impact of a local flood, and recommended that the council 'impose an embargo on granting planning permission for any development which would add to the flow in the surface water system' until a sewer scheme was completed.[30] Walker really was the archetypal local man, hearing concerns about fly tipping, tree preservation and a new car wash. He was a natural NIMBY ('Not In My Back Yard'), and upon learning of a plan to renew the Heathrow/Gatwick helicopter licence, leapt into action.[31] His committee, as 'a matter of urgency', resolved to alert local residents and conservation organisations about the impending noise pollution, and decreed that 'a Press release should be issued as soon as possible'. While he cared deeply about the people of Surrey getting a good nights sleep, he would not be so considerate about what his company did in other parts of the world.

This Venture Might Be Privatised

By 1983, with Tamil attacks increasing, Sri Lanka's president was once again 'baffled' about how to handle the situation.[32]

As he had done four years earlier, when MI5 had come to visit, Jayewardene turned to the British High Commission for help. The staff from the old colonial power had not gone far. They were stationed across the road from the prime minister's house in an ugly 1960s concrete block. Located on the Colombo sea front, it had no shelter from the Indian Ocean's westerly winds, which flung salt at the building and slowly rusted the reinforcing steel bars in the concrete, eating the building from the inside.

Despite these architectural difficulties, the team kept a close eye on the escalating civil strife. Britain's defence attaché, by now a Royal Marine lieutenant colonel, Ewan Sale, had just returned from a visit to Jaffna. He had spent a few days in plain clothes wandering around the Tamil city with a Sri Lankan army captain, who was armed with a loaded 9 mm pistol stuck down the back of his trousers.[33] Sale found that police stations were guarded with six foot barbed wire fences, concrete pill boxes, sandbags and sentries brandishing pump-action shotguns.

All of this was because of a group of militants that numbered no more than 50, including those already in custody, according to the local commander. This small movement was itself split between three rival Tamil factions. However, most of the police and soldiers stationed in Jaffna did not speak Tamil, and effectively resembled a foreign force. Police morale was low and they estimated that 90 per cent of Tamils in Jaffna wanted independence. In an attempt to fight back, the police had recently formed a 75-man commando unit, which underwent less than three months training in basic infantry tactics and jungle fighting. However, these police commandos were the exception, and the army expected the vast majority of police to concentrate on mundane matters such as traffic control rather than counter-insurgency. The chief constable grumbled that this rivalry between the army and police 'for supremacy in the counter-terrorist field was the greatest stumbling block'.[34]

Sri Lankan police and soldiers needed to start working together, and fast. Britain's Foreign Office hastily dispatched a police adviser on a special week-long visit to Sri Lanka in March 1983. The adviser proceeded to explain how counter-terrorism worked in Northern Ireland, emphasising that the police took primacy over the army wherever possible.[35] The British adviser specifically 'commended the lessons to be learned from the Morton Report', referring back to Jack Morton's visit to Sri Lanka. He also suggested that another MI5 officer visit Sri Lanka to further the aims of the Morton Report and review the structure of Special Branch.[36]

The Sri Lankan police chief welcomed this advice, and hoped to speak to senior Home Office officials about 'the role of police in internal security vis-à-vis the military'.[37] The trip continued in this vein, with the British police adviser repeatedly telling the Sri Lankans how things were done in Northern Ireland – where an elite police unit had recently taken over sensitive operations from the SAS. He met the cabinet secretary, who was so impressed with British counter-insurgency strategy that he asked 'whether a senior policeman could visit Northern Ireland and sit in with officers on the job'.[38] A month after the adviser's trip, Sri Lankan diplomats wrote to the British police with a new list of training requirements. They continued to want help with routine police work like traffic control and dog handling, but a new item stood out starkly at the top of the list: paramilitary training for counter-insurgency operations.[39] After months of intensive advice from the British government, the Sri Lankans now regarded paramilitary operations as a suitable role for policemen. In other words, Sri Lanka's police wanted the capability to carry out special forces-style operations, using intelligence gleaned from the surveillance apparatus recommended by Jack Morton.

However, the Foreign Office appeared slightly taken aback by the police's sudden enthusiasm for these more martial methods of law and order. A diplomat noted that the courses were 'of some political sensitivity and Tamil extremists in Sri Lanka could be expected to complain bitterly that HMG [Britain] was assisting in the training of the Sinhalese authorities, in order that they could continue their policies of "repression" of the legitimate rights and aspirations of the Tamil people in the country'. Yet these concerns appeared secondary. 'As you know,' he continued, 'we should like to help the Sri Lankan Government (discreetly) as much as we can with these courses.'[40] Having drawn up a counter-insurgency blueprint for Sri Lanka's police, the British authorities would stop short of implementing it directly. A more discreet strategy was preferred.

* * *

Sri Lanka's spring of 1983 was brutal and bloody, but worse was to come. The leaders of *Gandhiyam*, a respected Tamil humanitarian organisation supported by British charities such as Oxfam and Christian Aid, were arrested by police and allegedly tortured by the army to obtain false confessions that they were working with militants. Their offices were then destroyed by the army, along with many of the refugee resettlement schemes they had built up for Tamils over the previous six years.[41] This was not an isolated incident. The army torched Tamil properties across the island's north-east, from Jaffna to Vavuniya. Sinhalese mobs also went on the rampage in Trincomalee, and Tamil Tiger calls for boycotts of local elections were gaining a groundswell of support. The results of inquests and post-mortems began to filter into the Tamil conscience, confirming that Tamil youths had met violent deaths at the hands of the army. In response to the rising resistance, emergency regulations were passed giving the army the power to both kill and bury without a post-mortem or inquest, to protect the 'morale' of the security forces. Thereafter, a Jaffna Tamil youth, Sabaratnam Palanivel, was allegedly dragged into an army camp and run over by a truck, 'smashing his skull and flattening his body'.[42]

Despite these killings, Whitehall gave two senior Sri Lankan police officers a VIP tour of the UK in June 1983.[43] They were sent on an MI5 counter-terrorism course, and invited to visit the Metropolitan Police Special Branch to discuss 'the activities of organisations based in the UK agitating for a Separate State for Tamils in Sri Lanka'.[44] A senior civil servant at the Northern Ireland Office even arranged for the Sri Lankan guests to visit Belfast, 'to see at first hand the roles of the police and army in counter-terrorist operations'.[45]

Shortly after this trip to Belfast, the violence in Sri Lanka crossed a threshold. On 22 July 1983, three Tamil girls were allegedly abducted by the army in Jaffna, and rumours spread that one had been raped, while another had taken her own life. Retaliation by the young Tamil militants was swift and deadly: bombing an army truck the following day, killing 13 soldiers. The cycle of violence was spinning out of control. Troops in Jaffna killed 30 people that day in response, including six schoolchildren. News of the carnage reached the capital Colombo, where the Sinhalese majority were determined to avenge the death of their 13 soldiers. In the words of one Tamil author:

> From 24 July, the worst ever anti-Tamil rioting started. Hundreds of Tamils were killed, hundreds of Tamil homes and shops were looted and burnt. Despite the declaration of an all day and night curfew, looting and burning continued for several days following in the city, quite often in the presence of security forces. ... The Tamil people fled from their homes to various refugee camps, some of which came under attack by the Sinhalese mobs. At the time of writing, there are over 75,000 Tamil refugees in several camps in Colombo ... death and destruction have become the only things not denied to the Tamils.[46]

The British files are almost more scathing about the July 1983 riots than the Tamils were themselves. 'There is no doubt that elements of the army took an active part in the rioting, both in uniform and civilian clothes', commented Britain's defence attaché.[47] 'The Army is at present convincing itself that it did a good job during the riots – what blatant self-deception.' The problems were not only confined to that bloody week in July. Troops in the north were poorly trained and lacked discipline, a malaise which he felt contributed to 'the chronic hate the people of Jaffna have for the army'. However, more training was not necessarily the answer. 'In my opinion', Britain's defence attaché said, 'the equipment the army has is adequate, their training could be improved with guidance and patience, but their inherent indiscipline (aggravated by the dislike the almost wholly Sinhalese army has for the Jaffna Tamils) cannot be corrected and will therefore continue to be an Achilles heel'.[48] In time, this would prove to be a prescient remark.

Meanwhile, Sri Lanka's embattled president went on a charm offensive. He hosted a *Sunday Times* journalist at his country residence, the King's Pavilion, a lavish hill-country mansion which had been used by British governors. The journalist, Anthony Mascarenhas, had famously exposed the Pakistani army's atrocities in East Bengal during the Bangladesh war of independence. But he seemed to be taken in by the president's lavish surroundings, and hardly mentioned the fresh atrocities in Colombo or Jaffna. Instead he gushed: 'Dazzling displays of scarlet anthuriams flared from several vases; on one wall, starkly incongruous, hung a Japanese samurai sword, like a rip in an exotic canvas.' In this intimate setting, the president felt comfortable enough to hint at the impact of Britain's counter-insurgency advice his government had received over the previous months. 'A full-scale operation against the terrorists [was] due to start soon', he wrote. 'The president disclosed that a special force, mainly consisting of commandos but with some police, was being trained for the purpose.'[49]

While Jayewardene was busy charming foreign reporters, he made sure to appoint a ruthless defence secretary to crack down on the Tamils. General Attygalle was a veteran of the 1971 JVP uprising, where he was prepared to destroy entire villages.

Returning to the fore in 1983, the general was 'deeply disturbed' at his army's lack of training to deal with 'terrorism'. He asked the British envoy to send five soldiers to coach 150 members of his army in counter-terrorism. The diplomat tactfully suggested that a small number of Sri Lankan soldiers visit the UK for training there instead. He was privately concerned about Britain being seen to become more involved in the communal conflict, fearing 'unacceptable political consequences not only in UK but in India'.[50] Thousands of Tamil refugees had fled across the Palk Strait into the Tamil-speaking southern Indian state of Tamil Nadu, and the central government in Delhi led by Indira Gandhi was conscious that the Sri Lankan conflict was having domestic ramifications. Whitehall had to be mindful of India's concerns. Delhi, the regional superpower, was a major customer of British weapons, on a par with Saudi Arabia and the US. A trade embargo on UK–India arms sales would jeopardise jobs in Britain. Diplomats would have to tread a tightrope, supporting Sri Lanka sufficiently to defeat Tamil rebels without appearing in the eyes of India to be fuelling a refugee crisis into its southern state or meddling too much along its border.

Mindful of these geopolitical complications, Britain's most senior diplomat, Sir Antony Acland KCMG KCVO (later provost of Eton College) personally considered Sri Lanka's plea for UK military training. Acland was then the Foreign Office's permanent undersecretary, the top mandarin, trusted with liaising between the diplomats and the spies at MI6. The 53-year-old was from a military family, educated at Eton before reading PPE at Oxford. A career civil servant, he was a high-flier and establishment insider. He wrote a telegram stressing that 'this needs careful assessment' because of the 'Indian angle'. Indira Gandhi's personal envoy on Sri Lanka, Gopalaswami Parthasarathy, had told him at a meeting in New York that 'UK training of Sri Lankan security forces would *not* be helpful'. Still, Britain's top diplomat worried that 'we are not being very helpful to a friendly Commonwealth government [Sri Lanka] in difficulties'.[51]

Whitehall 'urgently' considered the request for military training. Britain was the only country Sri Lanka had asked for help, which added to the pressure.[52] The Foreign Office had an 'open mind' and said 'we presume that what is required is not the specialised counter-terrorist operations which, for example, the SAS can provide, but more general training in how to deal with terrorist groups operating within a particular community, a situation somewhat akin to that in Northern Ireland'.[53] The Ministry of Defence estimated that sending five soldiers to Sri Lanka for a six-month training course would cost £75,000. However, the department was concerned that the UK 'could face considerable embarrassment if our men became associated with ill-treatment of the Tamil minority', and that if the Sri Lankans wanted training in riot control that was beyond the British army's training remit.[54]

As the back room discussions rumbled on, the pressure on Britain to help was mounting. As well as the request for army training, Sri Lanka's new director of intelligence, Merril Gunaratne, wrote to Lieutenant Colonel Ewan Sale, the defence attaché at the British High Commission. Gunaratne was an Anglophile, who had attended several Special Branch courses in the UK, which he felt were of 'considerable value'.[55] He therefore enquired as to whether the training could be revived for some of his officers the following year. Gunaratne was well versed in the recommendations of the Morton Report, and although he never met its author, he was brought in after the 1983 riots to implement its recommendations.

At precisely the same time as Sri Lanka's new generation of spies were turning their attention to the Morton Report, the document was also being studied by senior British officials. The Foreign Office briefed Margaret Thatcher's special advisor, Sir Antony Parsons, on 'the background to the Morton Report', which seemed to dominate their approach to training Sri Lankan forces in counter-insurgency. (Parsons had become heavily involved in dealing with Colombo's request for military assistance after

Indira Gandhi's personal envoy on Sri Lanka, Gopalaswami Parthasarathy, called on him in Whitehall.)

Parsons was no stranger to rebellions. A decorated Second World War hero, who read Oriental languages at Oxford, he spent years working as a diplomat throughout the Middle East, helping to prop up pro-British rulers. Parsons was largely responsible for decades of repression in Bahrain, the tiny Gulf island north-west of Oman. He was the most senior colonial administrator (quaintly called the 'political resident') in 1966, as Britain prepared to hand back power to the local royal family, led by Sheikh Khalifa. Bahrain had just been rocked by a pro-independence uprising, and Parsons was plotting how to prevent Britain losing its grip on the strategically located oil-rich territory. He persuaded the Sheikh to appoint an experienced colonial policeman, Colonel Ian Henderson, as his head of Special Branch, and give him a free hand to reorganise it into an efficient, modern covert surveillance 'anti-terrorist' organisation.

Henderson was fresh from Kenya, where he played a leading role in suppressing the Mau Mau uprising – capturing its leader, who was hanged. In his new posting, Henderson became known as the 'butcher of Bahrain' for the savage torture of political prisoners that he oversaw. Parsons supervised Henderson for the first years of his tenure in Bahrain, from where he also meddled extensively in Oman's affairs. Parson's racism is evident from the British records, where he described the Omanis as 'a pretty jungly people'.[56] Parson's constantly fretted that the conflict had the potential to become a 'mini-Vietnam' for Britain. One of his key roles in the conflict was to help keep the Shah of Iran, another deeply repressive British ally in the region, supplying military support for the Sultan of Oman. Parson's close contact with the Shah made him the ideal candidate to become Britain's ambassador in Tehran in 1977, until the monarch was overthrown in the 1979 revolution – something that took Parsons by surprise, probably because he was so accustomed to being able to keep British client rulers in power.

So when, on 20 September 1983, the Foreign Office briefed Parsons on the Morton Report and the Tamil uprising, it was a profound moment. Parsons would have intimately understood Morton's perspective on counter-insurgency, and immediately set about trying to implement it. And he would also have grasped the geopolitical dimensions of Britain being seen to be too involved in Sri Lanka, fresh from his meeting with Indira Gandhi's envoy. At this moment, the files show that it was Parsons who 'wondered whether this venture might be privatised'. Parsons, Thatcher's special adviser, then explained that 'Major General (retired) Richard Clutterbuck was the leading UK expert on this kind of thing … he had now retired and was available for private commissions'. This old soldier was actually Parsons' neighbour in Devon, and the men were extremely close. Parsons had 'invited him to Iran privately in the last throes of the Shah's regime and found him to be an excellent adviser. He was now the director of a number of private security organisations and was a responsible operator.'

This encounter is a fascinating snapshot of how the British establishment functioned at the height of Thatcherism. Far from being seen as nepotism or a conflict of interest, the prime minister's special adviser was allowed to create and put forward his neighbour for a military consultancy role with a foreign ally, in order to avoid the appearance of overt British involvement in suppressing an uprising overseas. Clutterbuck's credentials for the job was that he had guided Parsons through the Iranian Revolution, while the Shah's forces shot protesters daily. He was also the author of several books on counter-insurgency and had seen action in countless British post-colonial campaigns, from Palestine to Malaya. For the Foreign Office, Parsons' suggestion could not have been more fitting. 'If Major General Clutterbuck were prepared to take on an advisory mission to Sri Lanka on a private basis, this would avoid many of the political complications of an official training team at this stage', a relieved diplomat replied.[57] In this emerging post-colonial world order, which balked at direct UK intervention abroad, legions of former British officers could now enjoy a lucrative retirement carrying out Whitehall's shadow foreign policy, far from the glare of public scrutiny. There was no need to tell Parliament anything about this arrangement. And if anyone did complain about the presence of British security contractors in Sri Lanka, Whitehall could simply 'shrug our shoulders and say we had no power to stop it', as one opponent of the Diplock committee had envisaged some seven years earlier.[58]

*　*　*

The elaborate plan was relayed to Britain's envoy in Colombo, who would have to sell the idea to Sri Lanka's leadership. His colleagues said they were 'considering the possibility of the Sri Lankan government seeking advice from private British organisations or individuals specialising in counter-terrorism and counter-subversion. These might be able to provide general advice to the Sri Lankans, while avoiding any political difficulties for HMG.' Without irony, the Foreign Office stressed that it 'would obviously be helpful if any British assistance could be seen in the context of supporting a Sri Lankan policy of reconciliation rather than confrontation' – even as it was proposing to unleash a war veteran like Clutterbuck, who played to win on the battlefield rather than seek compromise through patient negotiations.[59]

The matter was discussed in the UK at ministerial level on 7 October 1983. At least two offers of assistance were agreed. A handful of senior Sri Lankan military and police officers could visit London to hold discussions with police, including Special Branch, on the topics of intelligence, public order and community policing. 'We would wish to ensure strict confidentiality. There would be no publicity … discussions being held in secure premises', officials whispered.[60] In addition, Whitehall had decided: 'General Clutterbuck's name might be put to the

Sri Lankans as a private consultant to follow up the Sri Lankan Government's request for a British army training team.'

The general had been so eager to secure the job that he hastily knocked together a handwritten version of his CV on a Sunday to catch the first post the following morning from the sleepy village of Thorverton, near Exeter. In his scrawled letter, the retired general listed his military service with the British army in classic colonial hotspots from Palestine to Malaya, Borneo to Thailand, Hong Kong to Oman. After leaving the British army, he had carved out a career for himself as an academic at Exeter University teaching students about political violence. His skills were in demand from governments, police forces, armies, intelligence agencies and business around the world, allowing him to deliver 500 lectures in the previous five years alone. He had also written nine books about counter-insurgency and visited 14 countries to advise on the political and security situations, as well as undertaking regular advisory visits to Northern Ireland. Most worryingly from a human rights perspective was his visit to South Africa, which was funded ('indirectly') by the apartheid government, something he noted could be 'sensitive'.[61] He lectured their National Intelligence Service, police and army. The apartheid regime's hard line defence minister, General Malan, whom Clutterbuck knew, 'had me flown up to the Namibia-Angola border for 2 or 3 days'. Namibia was then illegally occupied by South Africa, and Namibia independence fighters were sheltering in Angola.

For the potential work in Sri Lanka, General Clutterbuck wanted to partner with the managing director of Asset Protection International (API), 'a specialist firm with which he is connected'. Diplomats welcomed this arrangement, although they did 'not wish to appear to be sponsoring API and it should be presented as *his* idea'. Again, the motivation was purely presentational. The official explained that 'we should take care to keep [them] at arms length in view of political sensitivities'.[62] API did not regard this convoluted arrangement as unusual. Its managing director told the Foreign Office that his company 'had been set up for just this purpose – where HMG did not or could not become involved'.[63] Clutterbuck was more than content to go along with the 'deniable' nature of his proposed role, recommending that he be 'presented as an independent authority on security problems'. The general would keep the Foreign Office informed of whatever arrangement they arrived at, and 'There would, of course, be no question of involvement by HMG in such discussion'.[64]

With Clutterbuck and API lined up, the British envoy was instructed to seek an early meeting with Sri Lanka's president to deliver the news, and warned that it should be emphasised this was 'a purely private arrangement in which HMG would have no involvement'.[65] After almost two months since Sri Lanka asked the UK for counter-terrorism training, the British envoy could now deliver Whitehall's immaculately crafted response to the president and defence secretary. While the president was disappointed that a British military training team would not be visiting his island, he 'appreciated' the alternatives on offer. A visit to London for some of his senior security officials was welcomed, and the defence secretary was interested to hear about General Clutterbuck. The envoy gave him a brochure for API, a spare copy of which seemed to be readily to hand at the British High Commission. There was one slight curveball in this meticulously choreographed discussion. It transpired that Sri Lanka's defence secretary was 'already in touch with a British team now in Oman' and it was unclear if this was Clutterbuck's company or another firm. In any event, the Sri Lankan president was interested and wished to know if he could pass an invitation to Clutterbuck through the British envoy, so long as that degree of involvement was not too much.[66] The British authorities were content to act as the middleman on this occasion.

A few days after this meeting, it became clear that the Sri Lankans were actually considering a different company from API. The high commissioner explained: 'The reference to the British team in Oman is to KMS Ltd, a British security firm (well known to the Security Department)', who had hired the firm's staff for bodyguards at British embassies around the world. 'Colonel Jim Johnson (ex-SAS) of this company has recently visited Sri Lanka and is returning next week to make proposals for training the police in internal security, counter-terrorist operations and VIP security. Before he signs any agreement with them Attygalle [Sri Lanka's Defence Secretary] would be grateful for your opinion of KMS before the beginning of next week.' Clutterbuck and Asset Protection were being sidelined, and only if there was money left over would the Sri Lankans consider asking them to 'offer similar training to the army in the New Year'. Amid the sudden change of plans, a diplomat scrawled at the top of the telegram: 'KMS Ltd successor Saladin Security Ltd', referring to the company's sister firm.[67]

Whitehall's response to its own envoy was nothing short of bizarre. It responded: 'we would not wish to offer any advice on KMS. Neither we nor [the] Ministry of Defence know much about the firm. The Foreign Office used them for some years for certain VIP protection tasks until 1981 (when the MOD [Ministry of Defence] was able to take over this responsibility) and they appeared competent and well organised. But we have no knowledge of their capabilities on police training.' The next line of the telegram is censored; however, its description of Whitehall's relationship with KMS is baffling. The British high commissioner had already stated that KMS was 'well known' to the Foreign Office's Security Department, which had hired it to work in Lebanon, the Netherlands, Argentina, El Salvador, Uganda and Uruguay – as recently as August 1982. MI5 had said it was the only firm to be trusted for this kind of work. Its staff were all former SAS high-fliers, who knew about some of the army's most sensitive operations. One SAS veteran from the period told me that, 'When I was in the SAS, everyone knew about KMS.' Therefore, the claim in this telegram that the Foreign Office and Ministry of Defence did not 'know much' about the firm should be read with care and counter-balanced with what we already know about Whitehall's relationship with KMS.

The situation becomes more complex as the rival firm, API, had in fact told the Ministry of Defence that 'KMS were their agents in Oman', strongly suggesting a business link between the two mercenary companies. Given that Margaret Thatcher's personal aide, Parsons, was well aware of the Sri Lankan request for military training from the outset, and that he had well-established connections to Oman, it is plausible that as well as discussing the situation with his next-door neighbour, General Clutterbuck, he also spoke to his contacts in Oman. For instance, Parsons would have known Tim Landon, the White Sultan, who was an associate in KMS, from his time as a diplomat in the Gulf.

Moreover, if one accepts that Whitehall did not surreptitiously set up the new KMS contract in Sri Lanka, it must be noted that they did not explicitly tell Sri Lanka to avoid hiring the firm – despite the opportunity being presented to them. A former defence attaché at the British High Commission, who started working there in 1985 (a year after KMS had arrived), was firmly of the impression that KMS had gone to the British High Commission and told them of their plans and were given permission to carry on, although there is no record of such a meeting in the files. The allegation that someone from KMS had visited Sri Lanka prior to the contract starting is supported by the evidence that Jim Johnson travelled to Sri Lanka in the midst of this manoeuvring. Given the pervasive censorship, and the explicit comments about keeping parts of this scheme 'at arm's length', it is conceivable that the exact cable about his trip was weeded out. The Foreign Office destroyed over 20 files about Sri Lanka, including defence attaché papers, from the year this contract was awarded. Such conduct calls into question whether the department is truly committed to preserving an accurate record of events surrounding this extremely delicate outsourcing arrangement.

Whitehall's disingenuous message was duly relayed, with the British envoy telling Sri Lanka's defence secretary that 'we knew little about KMS, but on the other hand had heard nothing adverse about them'.[68] The Foreign Office also updated Clutterbuck, telling him that Colombo was considering a different firm. By 24 November 1983, the decision had been taken, and General Attygalle told the British high commissioner that they were going to employ KMS for police training. The telegram was curt, with no commentary. It was copied to several parts of Whitehall, including an organisation or individual whose identity is redacted – presumably one of the intelligence agencies or special forces.

Soon, the defence attaché had gleaned that a KMS training team, rather than a sole consultant such as Clutterbuck, was scheduled to arrive in January 1984 and stay in Sri Lanka for a year. 'Their task will be to train 600 picked ranks aged 23 to 28', he said, revealing the scale of the scheme.[69] Of course, the company was not delivering anything as tame as regular police training. Instead, their students would be equipped with weapons, webbing and radios, and schooled in VIP protection and presidential protection. The company's protégés would operate not only in the relative safety of the capital Colombo, but also in Tamil flashpoints like Jaffna and Batticaloa. Intelligence-gathering capabilities would be 'grafted on' to the unit, which was shaping up to be an elite commando team, redolent of the KMS-led Sultan's Special Force in Oman. Indeed, the new squad's name, the Special Task Force (STF), was eerily similar. The counter-insurgency blueprint repeatedly recommended by British advisers like Morton and his colleagues was finally being implemented by the maverick Colonel Johnson, without Whitehall having to be directly involved. In private hands, progress would be swift. Every twelve weeks, KMS promised to churn out a new company of 120 police commandos, drawn from the Sinhalese community. The squad was unlikely 'to include any

Tamils since, so the General [Attygalle] claimed, it was difficult to get them to operate in Tamil areas'.

It was a recipe for ethnic warfare, but the training could not come soon enough in the eyes of Britain's deputy envoy to Colombo. He was worried by propaganda from the People's Liberation Organisation of Tamil Eelam (PLOTE), one of the most left-wing Tamil groups, and suspected that the Tigers were purchasing Kalashnikovs and training in India. 'We may then see some comparatively well equipped and comparatively well trained guerrillas making mincemeat of the army and the police, especially in the north', he said.[70] Although Whitehall was relieved to see the actual training of Sri Lankan forces deniably outsourced to a private company, it was still keen to ensure its ally was well equipped with British weaponry to put down the Tamils. In August 1983, just a month after the riots, the UK had agreed to supply riot control equipment to the Sri Lankan police and justified it on the grounds that it was to 'enable the security forces to protect the Tamils still in the south'.[71] Tear gas, CS grenades and launchers, stun grenades, bullet-proof vests, helmets and some 20,000 rubber bullets were all approved for export. Only electric shock batons were blocked because of fears they might be used for torture in light of a recent report by Amnesty International.

Months later, Colombo wanted more. The UK was asked to supply an array of armoured personnel carriers from manufacturers in Northern Ireland (Short Brothers Limited) and Wales (Hotspur Armoured Products Limited).[72] Some of these vehicles would end up being used by the KMS-trained STF. The Foreign Office approved the exports even though it said supplying military equipment to Sri Lanka was now more sensitive after Colombo had admitted that 'some undisciplined elements of the security forces did kill Tamils in the north of the country during the rioting'. Diplomats accepted that supplying arms 'could therefore be construed as support for repressive and undisciplined security forces'. However, the Foreign Office reasoned creatively that these vehicles would allow the security forces to 'deal with Sinhalese extremists on the streets and to protect the lives and property of the minority'. Whitehall reached this conclusion despite knowing that their 'main day-to-day role would presumably be patrolling, particularly in the Tamil north'. Some of the armoured vehicles Britain sold were equipped with 76 mm guns, effectively an artillery piece. The stage was now set for the next phase of this increasingly bloody war. Sri Lankan forces would fight with British weapons and be mentored by British mercenaries, but the British government itself would keep its hands clean. Having written the script, it could now sit backstage and watch the carnage unfold. For KMS, with its dwindling number of contracts, this profitable new war could not have come at a better time. No longer would it need to depend on official contracts with British embassies – there was a more lucrative market emerging in countries where the UK authorities did not wish to appear involved. And before long Whitehall would not be the only Western government to appreciate such a discreet service. Soon, there would be another powerful country that wanted to influence a foreign conflict, without appearing too deeply embroiled.

5

Oliver North's British Mercenary

Nicaragua, 1979

With Tamil and Dhufari revolutionaries waging guerrilla wars in Sri Lanka and Oman, some of their Central American comrades achieved a remarkable change in fortune. In the summer of 1979, the Sandinistas, an armed left-wing movement, toppled Nicaraguan strongman Anastasio Somoza Debayle, ending a family dictatorship that had spanned some four decades. 'Walls covered with graffiti, multicoloured spray-paint landscapes of slogans and the sayings of martyrs and designs appliquéd on doors and walls and pavements ... the revolutionary answer to the ubiquitous adverts of our lands', wrote one young socialist visitor from the UK, who travelled through Nicaragua at the dawn of this new political era.[1]

While many around the world celebrated, Margaret Thatcher was less enthusiastic about the fall of this right-wing tyrant. Initially, she was reluctant to even recognise Nicaragua's new government, because of its left-leaning agenda. Her foreign secretary tried to reassuringly describe the incoming rulers as a 'generally moderate, broad-based team with, so far, only one Sandinista member'. Other Western governments, most notably the US under President Jimmy Carter, were keen to recognise the new administration as early as possible – 'they hope thereby to avert the possibility of a gradual takeover and the threat of a Castro-style government on the Latin American mainland'.[2] After a week of hesitation, Thatcher grudgingly agreed to recognise the change of guard in Managua.

However, many of Somoza's loyal supporters simply could not countenance Nicaragua's new government. They launched their own guerrilla war against the Sandinistas, earning them the moniker 'Contras' or counter-revolutionaries (know officially as the Nicaraguan Democratic Force or FDN). Unlike in Sri Lanka and Oman, where the armed rebels were left-wingers, in Nicaragua the rebel movement was right wing. This important distinction transformed its international image in Western capitals, from terrorists to freedom fighters. The Contras' fortunes improved dramatically in 1981, when the Republican candidate Ronald Reagan won the US election. An avid free market capitalist, he authorised the Central Intelligence Agency (CIA) to covertly fund, train and arm the Contras to the tune of millions of dollars. The escalating conflict mired Nicaragua in turmoil and it was not until early November 1984 that the country was ready to hold presidential and legislative elections. The poll was designed to replace an ad hoc Council of State, set up shortly after the fall of Somoza and comprising trade unions, peasant organisations, business groups, professional associations, political parties and members of the women's movement. Despite the Sandinistas fielding this broad-based coalition, Thatcher's hostility towards Somoza's successors had not abated and she refused to send anyone from the UK to observe this landmark election, which she saw as a PR stunt by a Soviet-backed Marxist-Leninist regime.

Elsewhere, others were willing to be more open-minded about the Sandinistas. A month before Nicaragua went to the polls, President Reagan suffered a major setback when the US Congress voted to stop funding for efforts to overthrow the Sandinista government. Other institutions, such as Parliament's Human Rights Group in the UK, were eager to monitor the Nicaraguan poll. They sent an unofficial cross-party delegation, including the widely respected Labour politician Alf Dubs MP, as well as a Tory and a Liberal.[3] In the group's report on the proceedings, they said that the last five years of warfare had 'taken a heavy toll on Nicaragua, in lives and injuries and economic damage. The government has reported that in 1984, 600 civilians were killed by "Contras" and 1,400 were wounded, and that 1,000 Sandinista troops fell in combat. Economic damage in 1984 alone is estimated at $255 million.'

Against this backdrop of guerrilla war, the visiting British politicians were impressed by what they called the 'Sandinistas' positive achievements' which were 'predominately in the fields of health and education. In 1980, with the help of over 70,000 volunteer teachers, most of them secondary school pupils, the Sandinistas organised a Literacy Crusade, which was followed by adult education classes, again run by volunteers.' The reach of this movement into some of the remotest parts of Nicaragua was impressive. One socialist visitor wrote that when 'Travelling through the thick swathes of bird-filled verdant valleys and mountains ... we were met on the road by eighty truckloads of young brigadistas of the "People's Literacy Army"'.[4] The parliamentary group continued its praise for the left-wing movement: 'With inoculation and preventative health campaigns, they have made [a] significant impact on communicable and endemic diseases. Neither of these achievements would have been possible without the enthusiasm and involvement of the ordinary citizens of Nicaragua. They show what can be achieved by a poor country when people unite to work for common goals.'

In addition to health and education, the Sandinistas had made significant strides in land reform, redistributing 35 per cent of Nicaragua's arable land, much of it from the Somoza family empire. The visiting British parliamentarians did not see this policy as a socialist excess: 'The agrarian reform law protects efficient private producers and establishes no limit to the size of their holdings. Two-thirds of this land has been redistributed to individual farmers with the rest being divided between state-owned farms, which tend to be the largest agro-industrial enterprises with special management needs, and agricultural cooperatives.'

Turning to human rights, the British politicians found that there was no systematic government policy of torture, and that extrajudicial executions were 'extremely few in number'. They went on to reason that although such incidents were of deep concern, Nicaragua was still at war, and the scale of violations bore 'no comparison with the dictatorship which it replaced'. Certainly, by the time that Parliament's Human Rights Group arrived in the capital Managua to monitor the election in November 1984, they felt able to declare: 'In Nicaragua today, people do not put their lives at risk when they express political dissent.' There were no death squads as Reagan claimed, and the police were not carrying out disappearances. If British parliamentarians could recognise the achievements of the Sandinistas, then it was little wonder that oppressed people around the world saw this Nicaraguan movement as a powerful inspiration. In Sri Lanka, one of the more left-wing Tamil armed groups, PLOTE, was particularly enamoured with the Sandinistas. PLOTE devoted in-depth analysis of the Nicaraguan situation in its party magazine, *Spark*, spending page after page rebutting Reagan's denunciations of the Sandinistas.[5]

In the end, seven parties ran in Nicaragua's election, with four others abstaining. The boycott parties were all part of the Nicaraguan Democratic Coordinator (CDN), a right-wing grouping. They had published a long list of conditions the Sandinistas would have to agree to if the CDN was to participate in the election. The British MPs were unimpressed by the list, with its demands for a general amnesty for armed rebels and abolition of laws which infringed on private property rights. The UK delegation said the CDN's conditions 'could more properly be described as an electoral platform than as a basis for free elections'. On polling day, a quarter of the electorate abstained, and the Sandinistas led by Daniel Ortega won the election with 62 per cent of the total vote. The British observers said that the competing parties represented a wide range of ideologies, ranging from left to right, and that the CDN could have participated freely if it had chosen to. If anything, it was the US ambassador's visits to opposition leaders that constituted 'interference in the domestic political process of another country'. Ultimately, the British MPs said they were satisfied that the election was 'scrupulously fair'.

* * *

Washington, DC, December 1984

By now the Contras were virtually defeated by the Sandinistas, both militarily and politically. But the election results and reality on the ground meant nothing to the Sandinista's opponents in the White House, Langley and the Pentagon, who were determined to stop the spread of what they saw as communism in Central America – even if these communists were now democratically elected in a vote which British politicians regarded as free and fair. With a congressional ban on US funds for the Contras, the right-wing hawks had to formulate a more elaborate plan. The Contras would be supported through unofficial channels. Plausible deniability was key. A web of shell companies would be needed to administer the scheme without anyone being able to prove that the US government was involved. The proceeds of secret US arms sales to Iran, whose Ayatollahs were America's sworn enemies, would be diverted to bankroll the shell companies. What would one day become known as the Iran–Contra scandal was now underway.

As part of this intricate plan, the CIA director William Casey spoke to Reagan's navy secretary, John Lehman – a senior republican and former navy pilot. Lehman proceeded to ring an old friend from his days at university in Cambridge, England – a Mr David Walker. Five days later, the KMS man was at the Pentagon, standing in Lehman's office. At this point, Walker was still a serving Conservative councillor, although he was starting to tire of the role. He had attended twelve meetings that year, but had missed a series of events since October, possibly because the company's new contract in Sri Lanka was beginning to take off. He had stepped down from the town-planning committee, and taken up a slightly less onerous position on the council's environmental health panel. The highlight of the summer had been a 'very enjoyable' cricket match between council staff and the councillors, followed by an 'excellent buffet'. Walker had scrutinised the safety of a pelican crossing on a local high street. Life as a local politician was becoming increasingly pedestrian.

He could not resist the lure of his old life, and his old friend was about to plunge him straight back in, deeper than ever before. 'This weekend, at the request of Secretary John Lehman, I met with David Walker, a former British SAS officer who now heads two companies (KMS and SALADIN) which provide professional security services to foreign governments', US Marine Colonel Oliver North wrote in a top secret memo from early December 1984.[6] North was at the centre of the White House's secret war in Central America, and David Walker would become his brother-in-arms.[7] Although KMS had built its reputation by providing bodyguards and training troops, the company's work was about to shift towards a more direct role in fighting, bombing and eventually killing people, in highly questionable legal circumstances. US government files suggest that Walker had begun supporting the Contras sometime before the Nicaraguan poll, and continued to favour them despite the Sandinistas winning a democratic mandate. This reactionary approach towards Nicaraguan politics gave Walker common cause with Oliver North, who was taking a close interest in his company.

'In addition to the security services provided by KMS, this offshore (Jersey Islands) company also has professional military "trainers" available', North told Reagan's national security adviser, another former US marine, Robert 'Bud' McFarlane. 'Walker suggested that he would be interested in establishing an arrangement with the FDN for certain special operations expertise aimed particularly at destroying HIND helicopters. Walker quite accurately points out that the helicopters are more easily destroyed on the ground than in the air'. The next two lines of the memo, titled 'Assistance for the Nicaraguan Resistance', are redacted, but it is clear from the uncensored section that Walker envisaged sabotaging Nicaragua's fearsome Soviet-made Mi-24 helicopter gunships, known in military circles as HIND. This level of direct mercenary activity would be a step up from the training and advisory roles that KMS had provided in other conflicts thus far. But if their sabotage raids were successful, then the rewards could be enormous. President Reagan claimed to regard the aircraft's presence in Nicaragua as a serious escalation in the Cold War balance of power, marking a build-up of Soviet military hardware in America's so-called 'back yard'.

Despite the possible geopolitical implications, the Sandinistas were more interested in the helicopters' utility for domestic counter-insurgency purposes. Since the advent of manned flight, air power has become a decisive factor in almost all armed conflicts. David Walker knew this only too well. He had helped win the Battle of Mirbat in Dhufar by using a helicopter to outflank the revolutionaries and pin them down in a killing zone. In Nicaragua, the tables were turned and David Walker found himself on the side of the rebels, who many regarded as little more than terrorists. To make money from this conflict, Walker would have to show that he was equally capable of thinking like an insurgent in the mountains of Nicaragua as he was able to direct a counter-insurgency strategy in the Omani desert. Immediately, he reached the same conclusion as many guerrilla commanders had done before him – the best option was to destroy the helicopters before they took off, by penetrating deep behind enemy lines, infiltrating the airbase perimeter, sabotaging the helicopters and escaping as quickly as possible – a classic special forces mission, straight out of the SAS playbook. Oliver North was clearly impressed by Walker's grasp of covert military strategy, and concluded his top secret memo by informing President Reagan's national security adviser: 'Unless otherwise directed, Walker will be introduced to Calero [a Contra leader] and efforts will be made to defray the cost of Walker's operations from other than Calero's limited assets.' Adolfo Calero was a former Coca-Cola executive, who did not care where the money came from for his guerrilla war. He once said, 'When you're in the desert and you're dying of thirst, you don't ask if the water they are giving you is Schweppes or Perrier. You just drink the damn thing.'

Just weeks after Walker's transatlantic trip, Margaret Thatcher was preparing to visit President Reagan in Washington. North excitedly sent the national security adviser another top secret memo, trying to make the most of the forthcoming visit and shedding more light on what he had discussed with David Walker. Not only had Walker 'indicated a willingness to assist in special mission planning and training for the Nicaraguan resistance', but the Englishman had also told North that 'BLOWPIPE surface-to-air missiles may be available in Chile for use by the FDN in dealing with the HIND helicopters'.[8] If the Contras could not sabotage the helicopters while they were on the ground, then the guerrillas would be vulnerable to aerial bombardment. The only way to survive this type of assault was with shoulder-launched rockets. The best Britain's military had at that time was the 'Blowpipe' model, which was used in the Falklands War with dubious results. One brigadier compared it to 'trying to shoot pheasants with a drainpipe'. For the Contras though, this technology was a glimmer of hope. North's memo shows that as soon as Calero was tipped off about the nearest stockpile, the Contra leader 'proceeded immediately to Chile'. There, he found that the Chilean army,

have possession of large quantities of BLOWPIPE missiles and that they are willing to make 48 missiles and five to eight launchers available to the FDN. There would be no charge for the missiles, but [redaction – looks like a Chilean official] did ask $15k each for the launchers. ... Training on the weapons system was also offered for up to ten three-man teams from

the FDN on a no-cost basis. Calero will dispatch the trainees
to Chile on December 23.

However, there was one snag. The Blowpipe missiles had been
manufactured in the UK, and sold to Chile with an end-user
export licence. To sell on the missiles to a different user, the
Chilean army had said it 'would need British approval for such
a transfer'. Colonel North felt that Thatcher's pre-Christmas
visit to Washington in two days' time, 22 December, 'offered an
opportunity to address the issue'. This ambitious colonel recommended that 'the President raise the issue privately with the
Prime Minister'. Colonel North then went on to say that Thatcher
had 'available a non-reportable contingency fund', a mysterious
pot of money which presumably could help fund the Contras.

To expedite his scheme, the colonel drafted a memo for his
superior, the White House's National Security Adviser Robert
McFarlane, to send to President Reagan, warning that it 'may
be the best opportunity we have to obtain such support for the
resistance. Given recent actions in Nicaragua, we may not have
another chance if we wait.' Although McFarlane denies ever for-
warding the memo, the draft is instructive for understanding
North's game plan, noting that the president's private meeting
with the prime minister 'would offer an opportunity to sound
out HMG on steps they could take to assist the Nicaraguan resistance'. The next three lines are censored, but it goes on to advise
the president to 'very privately express approbation for anything
HMG is willing to do in support of the cause. The Chilean government has not yet approached HMG on transferring Blowpipe.'
Thatcher would need to be 'fully aware of the constraints' on
the CIA and Pentagon, as technically they were banned from
funding the Contras. To sway the balance, the hawks suggested,
'We would be willing to offer her a discreet briefing on the Soviet
Bloc build-up in Nicaragua and its consequences for NATO if
the resistance is forced to go it alone without outside support.'[9]

Although McFarlane denies that he ever asked 'the President to intercede with any person for the obtaining of Blowpipes for the Contras', he attended a working lunch at Camp David between Thatcher and Reagan just after North's memo, where Nicaragua was discussed at least in some capacity. Downing Street papers from the meeting show that Reagan stressed the size of Nicaragua's military, and Thatcher took the bait, observing that 'the Soviets now seemed to be sending additional ships with arms'. Reagan readily agreed, adding that 'one of these ships had contained MIG [Soviet] aircraft'. He also said it was hard to detect the 'precise cargoes these ships carry' because of gaps in their surveillance. Thatcher called the situation 'very worrying'.[10]

Despite this meeting, when questioned by reporters the UK government would later deny that it gave permission for Chile to divert its stockpile of Blowpipe missiles to the Contras. And indeed, the Contra leader Adolfo Calero disappointedly told Colonel North on 3 January 1985: 'Blowpipe deal is off'.[11] However, there was still some cause for optimism. In a letter from February 1985, Colonel North wrote to Calero about the 'best news of all', saying that 'a sum in excess of $20 million will be deposited in the usual account'. He explained: 'This new money will provide great flexibility we have not enjoyed to date. I would urge you to make use of some of it for my British friend and his services for special operations', in what appears to be an oblique reference to David Walker, who he pledged to produce at the end of February. 'You and I both recognize his value and limitations', North murmured, 'I will find out how much he is getting and let you know, but it seems as though something should be set aside for this purpose.'[12] The source of this extra $20 million is unclear, but the information was so politically sensitive that Colonel North instructed the Contra leader to 'destroy this letter after reading. ... Please do *not* in any way *make* anyone aware of the deposit.' He continued to emphasise the need for utmost secrecy. 'Too much is becoming known by too many people', North lamented. 'We need to make sure that this new financing does not become known.' The reason for this discretion was highly political. 'The Congress must believe that there continues to be an urgent need for funding', North explained, before signing off using his codename – 'Warm regards, Steelhammer.'

In the body of the letter, North revealed how vulnerable the Contra forces were to any attacks by the Nicaraguan army, ordering Calero to disperse his forces around the high ground 'so that they are not caught in the firestorm as the Sandinistas intend'. These precautions were necessary to buy time while new recruits were trained up, and logistics strengthened. The anxious colonel went on to stress:

> Most important is saving the force from what I believe will be a serious effort to destroy it in the next few weeks. While I know it hurts to hide, now is the time to do it. While they are hiding, the man who is carrying this message can start the regular resupply process. I believe it would be wise to dedicate as much as $9–10M for nothing but logistics.

If Calero could survive the short-term onslaught, then the tide would begin to turn. Once the logistics and training were completed, North envisaged that opportunities would arise for 'hitting them [Nicaraguan government troops] hard as they phase down in frustration from their current operations and striking at selected strategic targets with your enhanced capability'. Colonel North's cryptic reference to 'strategic targets' and 'enhanced capability' would soon become clear.

A Businesslike Officer

Katukurunda lies 30 miles south of Colombo, along the Indian Ocean coastline. My grandfather remembers it as a Royal Navy airstrip during the Second World War. A runway was carved out of the jungle, from which British naval pilots flew sorties to hunt Japanese submarines. Black-and-white photos in the Imperial

War Museum archive show young British sailors, topless, relaxing under a cool colonial veranda sipping tea served to them by a local Ceylonese man, who is wearing a sarong and staring nervously at the camera.[13]

The base was closed down after the war, as the British sailors slowly trickled home. Forty years later, and there is a different set of photos from Katukurunda. These are not to be found in any official British archive. Again they show healthy young white men with military haircuts, but this time they are wearing shirts and ties, although it is hardly a uniform. They look more like businessmen. Some smile contently at the camera. One white man with a clipboard looks slightly bemused. A few, however, are practically snarling at the camera. Sitting among them is a handful of Sinhalese policemen in brand new uniforms. This is a rare photo of the STF with its British instructors, from KMS.

The STF training camp at Katukurunda was up and running by January 1984, after all the bureaucratic chicanery of the previous year. Despite winning this important new contract, some KMS managers were initially unphased by the company's operations. David Walker attended two Elmbridge council meetings that month, where he heard about an application for a 'license for the sale of ice cream'.[14] Meanwhile, his partners in Sri Lanka were wasting no time at all in the lush tropical climate, where they had been busily building a replica of the SAS base at Hereford. I know this because in 2018, two of my colleagues, Lou Macnamara and Dr Rachel Seoighe, who were assisting me in making a documentary about KMS, were unexpectedly granted permission to visit the STF camp. After months of trying to gain access to this formidable police unit, the pair were treated to a one-day guided tour, in which they were allowed to film the base, interview several current and former staff, and inspect some archival material. From their footage of that rare glimpse behind the wire, it is possible to see how a carbon copy of the SAS camp at Hereford was transplanted to Katukurunda courtesy of KMS. A close-quarter battle house, where SAS men practise hostage rescue, is replicated there. An abseiling tower, with a helicopter ski at the top and gaps for window frames below, is also present.

As my colleagues toured the camp, they were treated to several live fire exercises by their eager hosts. In one display, STF officers dashed between stencils of women and children who were wearing 1950s English fashion, erected among flower pots and juxtaposed with more menacing cut-outs of masked gunmen. They fired at the imaginary hostage-takers with their Heckler and Koch MP5s, before switching to their side arms, 9 mm Browning pistols – exactly as the SAS would do. Huge earth embankments, several stories high, demarcated a larger shooting area, a long linear strip, with a Buddhist stupa tucked safely away in the distance. Camouflaged figures, armed with M-16 assault rifles, stealthily prowled between the trees. They broke cover and opened fire, with a machine gunner joining in. Scores of motocross bikes came charging in behind them, delivering more men to the mock battlefield. Smoke flares billowed purple into the sky.

Little had changed since the KMS men designed these first training exercises. One addition is a gymnastics routine, in which hundreds of young STF recruits formed human pyramids and gyrated to Queen and Michael Jackson. Aside from this bizarre spectacle, my colleagues continually found reminders of the past at Katukurunda. The chief instructor's office was one such throwback. An immaculately varnished wooden board hung from his wall, carefully listing the names of his predecessors. His name is at the bottom of a line of Sinhalese names, but two English names at the top are incongruous:

Special Task Force training school chief instructors
1. Mr Ginger Ress, 10 February 1984 to 17 December 1985.
2. Mr Tom Hegan, 17 December 1985 to 8 February 1987.

For the first three years, the chief instructors were British, and their names are not just consigned to the past. Some of the staff at the camp have fond memories of these men. 'I joined the police department in 1984', Athula Daulagala, the director of the camp, told my colleagues. 'We observed the first training batch, trained by the KMS', he recalled. 'We were taken from the training school to work as the riot team, a civilian riot team performing the scenario based training... to act as the unruly crowd. So that was the first time the foreigners had trained the Special Task Force.' Smiling broadly in his spotless camouflage uniform, adorned with awards, the director recounted his earliest memories of the KMS staff. One of them taught him to use the M-16 assault rifle, another how to handle improvised explosive devices.[15] As he reminisces, an orderly in a white, colonial-style uniform is never far away in case tea or cake are called for.

He continued to describe in detail what training KMS provided: tactics, fieldcraft, navigation and communications. 'We are using the same voice procedure as the British army', Daulagala explained. KMS equipped the commandos with a cutting-edge encrypted radio system known as Cougar, which was used by British special forces and proved vital for Sri Lanka's counter-terrorism operations. Daulagala is not the only prominent STF officer still in service who was originally trained by KMS all those years ago. The commandant of the STF in 2019, Senior Deputy Inspector General Latiff, was part of the first cohort of trainees in 1984. Three and half decades on, the legacy of KMS is very much alive in Sri Lanka.

So who were the British names on the chief instructor's board? They are now ghost-like figures, who have vanished without a trace. Tom Hegan was probably not his real name, and the KMS men were known to use aliases to cover their tracks and conceal their location from IRA assassins. Ginger Rees, or Ress, was clearly a nickname, but one that was instantly recognised by a Northern Ireland veteran from that period who I consulted. Flicking through a battered old address book, my source confidently confirmed that Ginger Rees would have been Malcolm Rees, the SAS sergeant who was arrested by the Irish police in

1976 for illegally crossing the border with his heavily armed unit. Records show that after that debacle, he served another tour in Northern Ireland in 1981/2 where he won an MBE 'in recognition of his distinguished service'. He retired from the British army as a captain on 1 February 1984. A week later, he was in Sri Lanka working for KMS as the chief instructor at the STF training school. The man who made the map-reading error had moved seamlessly from Hereford to Katukurunda.

The Irish border fiasco was clearly not a barrier to prevent the men involved from finding lucrative work with a company like KMS. Sometime after Rees began work at Katukurunda, he was joined by another SAS veteran associated with that botched operation. His commander, Brian Baty, had eventually retired from the British army as a lieutenant colonel in October 1984 and was soon dispatched by KMS to oversee the company's operations in Sri Lanka. 'We even had the former commandant of the training wing of the SAS with us, Brian Baty, so we had experienced people', a retired STF chief has confirmed anonymously. 'Not green ones!' However, such battlefield experience came with risks attached, and Sri Lanka's defence secretary was particularly worried about the security of such a high-profile mercenary like Baty. His home in Hereford was being watched by a dangerous subversive, Peter Jordan, a retired teacher and dedicated revolutionary who had spent two years conducting reconnaissance on Baty and a host of British generals and spymasters at their remote rural homes. Jordan teamed up with a member of the Irish National Liberation Army, a republican group to the left of the IRA, to execute his plans. On Christmas Eve, 1984, Jordan arrived at a Liverpool pub to collect an important package. Unbeknown to him, his every move was being watched by dozens of Special Branch officers before they swooped on him. He was sentenced to 14 years in prison for plotting to blow up Baty's house.[16]

One of the earliest STF commanders recalled the dilemma: 'He was a target of the IRA in England at that time so then he came and served us here. To keep the anonymity we gave him the name of Ken Whyte.' By the time Baty, alias Whyte, arrived, the STF training camp was well established. However, his seniority impressed the Sri Lankan top brass. 'We had ones who had served in the Malaysian campaign and all that, so they could share their expertise and their knowledge', the former STF chief regaled, omitting mention of Baty's controversial experience in Northern Ireland. It seemed that the Sri Lankans were more interested in his experience of fighting in tropical conditions. Baty had run the SAS jungle warfare wing, and as a sergeant had won the prestigious Military Medal for an ambush in Malaysia. His team slayed six Indonesian soldiers who strayed into the former British colony. Army awards note that the success of that operation was due entirely to Baty's 'determined leadership and aggressive action'.[17]

Merril Gunaratne, Sri Lanka's former director general of intelligence, is one of the greatest admirers of Baty/Whyte. 'He was very good, a very professional guy, a very tough guy and the kind of guy who gave the necessary leadership in the emergence of the STF', Gunaratne told my team, as they intrepidly visited his home. The old spymaster credited Baty and his fellow mercenaries with playing a pivotal role in the war effort. 'The Special Task Force that we know of today owes its emergence and growth entirely to the KMS', Gunaratne marvelled. A keen cricketer and animal lover, and passionate vegetarian, Gunaratne had a ringside view of KMS operations. Nestled on his pristine cream sofa while household staff brought trays of spicy Sri Lankan pastries, he proudly told my colleagues how Sri Lanka's security forces scrambled to adjust to the new tempo of militancy after the 1983 riots.

Initially, Gunaratne helped set up the 75-man police commando unit in the north, which guarded police stations but could not carry out offensive missions. When he became director of intelligence in August 1983, the president's son, Ravi Jayawardene, told him the security forces needed a more professional unit that could 'play the LTTE at their own game'. Ravi wanted a clandestine unit that could use the element of surprise, and the only way to create one was to 'bring in foreign blood for training'. In Gunaratne's version of how KMS came to train the STF, he places particular emphasis on the role of President Jayawardene's son, suggesting that the company was trusted at the highest level of the Sri Lankan government. Under KMS's tutelage, and the patronage of the president's son, the STF rapidly became an elite unit, with better weapons and equipment than the rest of Sri Lanka's security forces.

As the summer of 1984 drew to a close, UK diplomats monitored the arrival of the STF's new weaponry in Sri Lanka. American M-16 and Belgian FN rifles, which used 5.56 mm calibre bullets, were being purchased and trialled by the new commando unit. One cable noted that the Sri Lankan government was 'concerned that these weapons cause much greater havoc to what they hit than the older .303 and 7.62 weapons'. The defence secretary, General Attygalle, felt: 'This is acceptable if a terrorist is hit, but not acceptable if an innocent citizen is hit.' In light of this, Britain's defence attaché asked the Ministry of Defence for advice on what alternatives were available that would not have such a 'devastating effect' on a human body. The type of ammunition was not the only issue causing concern:

> General Attygalle said that he was satisfied with the training being given by KMS to the STF although it was taking rather a long time. He also appeared slightly concerned by the fact that KMS was making each trainee fire 3,000 rounds of ammunition. Although costing rather a lot of money, General Attygalle acknowledged that this was the only way to give the trainee confidence and reasonable skill in using a weapon.[18]

A British official wrote an exclamation mark in the margin next to the number of bullets, but the training continued.[19] Within days of this telegram to London, the consequences of such a trigger-happy approach were becoming painfully apparent.

* * *

Joseph Rajaratnam is a 94-year-old retired maths teacher living in the northernmost tip of Sri Lanka, Point Pedro. When I arrived at his house on a blisteringly hot summer day, he was sitting on his porch waiting patiently. He led an ascetic existence: eating a spartan diet, fasting frequently and praying often at the local church. He welcomed me inside where the cool darkness felt like a sanctuary. The walls were painted an earthen red on which hung faded photos of his wedding day. He was now a widower, his wife having passed away seven years previously.

I had come to see him because in 1984 he worked at Hartley College, the oldest educational institute in the north, founded by an English missionary and scholar over a century ago. The school was renowned for producing engineers, doctors and scientists. Rajaratnam had spent over two decades teaching there, and in 1984 he was about to retire. The war was increasingly taking its toll on the school and his pupils were often harassed by the army. 'The parents are afraid of sending the children to college as long as the commandos are in the neighbouring compound', one journalist wrote at the time, and it appears that an STF camp may have been located next to the school.[20] The security forces wanted a heavy presence in Point Pedro because the town was a hotbed for the Tamil resistance, with the LTTE leader's village just along the sea front. Although Rajaratnam avoided politics, and focused on teaching mathematics, it would soon become impossible for him to ignore the war.

At lunchtime on 1 September 1984, an electronically controlled LTTE landmine exploded underneath an STF convoy two miles away from Point Pedro. Four STF men were killed and three seriously injured. The elite unit did not simply lick their wounds, they went on the rampage. 'Further reports indicate police retaliated by burning shops and other buildings in Point Pedro. There are also reports of civilian deaths. Reports vary from 6 or 10 civilians (government) to 18 (Tamil United Liberation Front)', a British telegram buzzed.[21] Hartley College was not spared and the library went up in flames.[22] 'There were more than 2,000 books, valuable books, everything burnt', Rajaratnam lamented, as he leafed through dusty school yearbooks to find a forlorn photo of the books burning.

This carnage at Point Pedro was the STF's first war crime. However, Sri Lanka's national security minister, the hawkish Lalith Athulathmudali, saw things a little differently. He told Britain's defence attaché that he was 'quite pleased with the performance of the KMS-trained Police Special Task Force: he was impressed by the way that they fought back at Tikkam [Point Pedro] and believes that the terrorists respect and fear them'.[23] British diplomats were slightly more circumspect. 'Elements of the STF have already been in action', the defence attaché noted in reference to the Point Pedro incident. 'But it is difficult to assess their performance. Unless, however, they prove themselves to be better disciplined than the Army, it will be a case of "out of the frying pan and into the fire".'[24]

There are no records available that would shed light on the reaction of KMS to the company's first brush with war crimes. The only indication of the company's attitude is that two days after the Point Pedro massacre, David Walker attended a Recreations and Amenities Committee meeting at Elmbridge Borough Council. There, he and his fellow councillors discussed how the local Elmbridge museum could assist with a travelling exhibition on wedding fashions. The committee duly resolved to 'make available specimens of wedding dresses and photographs from their collection for inclusion'.[25] They also discussed caretaking and cleaning services at Hersham village hall, as well as facilities for Walton hockey and cricket clubs. Naturally, no one suggested making a donation towards repairing the Hartley College library.

Instead, there should have been serious questions asked about the effectiveness of KMS training at this point. The regular Sri Lankan forces already had a long record of retaliating against innocent civilians when patrols were ambushed. The STF, however, was a brand new unit that had been intensively trained by some of the British army's most professional and disciplined alumni. As such, its influence should have moderated these beleaguered STF men in Point Pedro. Perhaps this particular collapse of discipline was merely an anomaly, and given more time the KMS training would begin to reign in the more hot-headed commandos. Or, more disconcertingly, perhaps the defence attaché's observation from twelve months earlier was completely accurate, when he lamented that the security forces' 'inherent indiscipline ... cannot be corrected' however much training they received.[26]

This warning had clearly fallen on deaf ears, for instead of halting any British assistance, Sri Lanka's defence secretary, General Don Attygalle, arrived in London a month after the Point Pedro massacre for discussions with KMS.[27] He also popped over to Baty's old stomping ground, Northern Ireland, to see some serving British personnel. There, the chief constable of the RUC, Sir John Hermon, 'entertained him to luncheon' on a Sunday.[28] The visit was kept secret. It was a particularly tense time in Belfast, with Hermon's own police commando unit under investigation for killing several unarmed men. The Sri Lankan general had sought out the RUC boss for advice on two pressing issues: 'anti-mine precautions for mobile security force patrols' and 'current tactics used by security forces in built up areas', being the same exact issues that the STF had recently faced in Point Pedro. There is no sign that his British hosts raised concerns with him about the most recent bloodshed. Instead, a civil servant noted patronisingly, 'In its small way this visit may have done some good. The General seems to have learned something from it.' Attygalle sent a letter to Hermon thanking him profusely for the 'useful advice and frank briefing' on the issues faced in Sri Lanka, claiming he had already implemented some of the 'excellent ideas' proffered in Belfast.[29] Although KMS was increasingly dominating the scene in Colombo, the country's security chiefs could still command an audience with serving British personnel, who were more than happy to entertain them – seemingly no matter what atrocities their forces committed.

* * *

The year 1984 had been a profitable one for KMS, seeding new contracts with Sri Lanka's police commandos and the Nicaraguan Contras. Still, there was scope for the business to expand before the year ended. On 28 December 1984, a retired British army colonel answered the telephone at his house in the countryside.[30] Colonel Jim Johnson, by then designated as the chairman of KMS Ltd, was having to deal with a business phone call in the supposedly quiet week between Boxing Day and New Year. For him, it was a time of year with much to celebrate. He had turned 60 just days before Christmas, and his future promised many more mysterious adventures in an already intriguing life.

Indeed, as this phone call was about to demonstrate, his company had a curious relationship with the British state. When Colonel Johnson answered the telephone, the voice that greeted him down the line would have been equally well spoken. The caller was a former Royal Marine, who had become a career diplomat. Sir (George) William Harding, known as 'Bill', had by now spent over three decades in the UK's Foreign Office. He would have already come across several of Johnson's KMS men – especially Malcolm Rees and Brian Baty, who now had senior positions in the company. Harding had run the Foreign Office's Irish department in 1976, when Rees and Baty were caught up in the SAS border-crossing fiasco. None of their careers had been hampered by this embarrassing diplomatic incident, however. Harding was now a deputy undersecretary of state at the Foreign Office, one of Britain's most senior diplomats, with responsibility for 'the Americas and Asia'. It was a vast geographic remit, and when he called Colonel Johnson, he had to make clear that he 'was speaking on instructions from the Secretary of State'.

It would transpire during this phone call that Britain's foreign secretary, Geoffrey Howe, wanted the KMS chairman to know that ministers 'would be very strongly opposed to any involvement in a combat role in the worsening security situation in Sri Lanka of the KMS personnel who were at present engaged in training the Sri Lankan police'. The reason for this concern was that British diplomats believed, perhaps in reference to the Point Pedro massacre or another atrocity, that the Sri Lankan security forces had 'effectively lost control' and a very dangerous situation was now developing. 'It was very important that his men should avoid being sucked into it', he warned Colonel Johnson. The KMS chairman 'took this in good part' and explained that his team were currently confined entirely to training the Sri Lankan police. However, he let slip that President Jayewardene's son Ravi and his national security minister, Lalith Athulathmudali, had both seemed 'desperately concerned at the way things were going' when they phoned him shortly before Christmas.

The next four lines of this telephone transcript are redacted, suggesting that something extremely sensitive was said next which Whitehall wishes to keep secret over three decades later. Did Johnson explain to one of Britain's most senior diplomats about Ravi's enthusiasm to develop SAS-style units to take the fight directly to the Tigers? Such a conversation is entirely plausible given that the fifth line reveals that the security minister had asked Colonel Johnson if his team could take on a combat role, 'pointing to the precedent in Oman, where KMS personnel had done exactly that'. This comment suggests that the KMS men in Dhufar with the Sultan's Special Forces were not merely confined to barracks. But the KMS chairman flatly refused to replicate this arrangement, 'not as a matter of principle, but because he regards the Sri Lankan army as totally out of control, with no leadership or staff backing, and thus a very dangerous fellow combatant. He remarked that they had so far lost their head that they had resorted to the wholesale massacre of women and children.'

The phone call made clear that one of the most senior KMS figures, Colonel Johnson, and one of the Foreign Office's top diplomats, Harding, both regarded the Sri Lankan security forces as a liability. And yet the solution they arrived at was to have profound implications. Colonel Johnson thought it might be useful to set up a modern battle school to train Sri Lankan soldiers, 'who could then take on the guerrillas at their own game', a turn of phrase that the president's son would have recognised. The colonel assumed that the foreign secretary had no objection to KMS currently coaching the police, and therefore surely would not mind if the company extended this training role to encompass the army as well. Harding did not think this would cause any problems, 'so long as there was no question of any participation whatsoever in active combat'.

With that, Margaret Thatcher's British government had given its approval, via Harding, for KMS to deepen its involvement in the Sri Lankan conflict. Not only was KMS training the police, they would now begin training the Sri Lankan army as well – a branch of the security forces that KMS knew was responsible for war crimes. Ironically, although Harding had made the phone call with the avowed intention of limiting the company's role in Sri Lanka, he had readily given KMS the green light to take on this new venture. He justified it with a splash of chauvinism, explaining in the transcript that Colonel Johnson was afraid that 'the Sri Lankan Government were so desperate that they would go into the open market in search of mercenaries that might end up with an ill-assorted and undisciplined bunch of Americans, French, Belgians and others, who would make a bad situation worse'. The implication was that it was far better for well-disciplined British mercenaries to train the Sri Lankan army, before the dastardly Yanks saw an opportunity.

The phone call ended with Harding suggesting that the men speak again about the situation at length in the New Year. Colonel Johnson readily agreed, and promised to ring the Foreign Office in mid-January when he returned from a short trip to Oman (presumably where he was attending to the Sultan's Special Forces). Harding was clearly impressed by the KMS chairman, commenting: 'As one might expect, Colonel Johnson sounded to be a brisk, efficient, and businesslike officer. He seemed very ready to listen to advice and willing to cooperate over a worsening crisis which is obviously giving him a good deal of worry.' The final six lines of the document are redacted, so we do not know what other observations he made. Harding sent the memo of his phone call to a more junior colleague in the Foreign Office's South Asia Department, Bruce Cleghorn. Although Sir William and Colonel Johnson have both since died, Cleghorn, after a successful career in the Foreign Office, rose to become an ambassador. He now works for the department in a censorship role, carefully vetting old telegrams before they are released to the National Archives under the 30-year rule. Whatever else Harding said in this phone call, Cleghorn and his fellow censors have made certain it will not be released to the public. Despite these attempts to hide aspects of the past, there is enough evidence to show that when it mattered most, a senior representative from Thatcher's government gave KMS approval to train Sri Lanka's army, which was in the midst of carrying out war crimes.

What happened in the months after their New Year's Eve phone call is much harder to piece together. The Foreign Office has destroyed dozens of files about Sri Lanka from 1985, including one that was ominously titled 'UK military assistance to Sri Lanka'. The department has also tried to block the publication of two files from 1985 about the 'Involvement of UK companies training Sri Lankan security forces', although it eventually disclosed parts of the files in response to a freedom of information request I filed.[31] What they show is that the minutes of Harding's phone call with Colonel Johnson were seen by the foreign secretary himself, Geoffrey Howe, who was rather more concerned about the potential ramifications than his buccaneering colleague. Howe wanted to know the legal and political position on mercenaries, apparently oblivious to the legislative black hole left by Whitehall's lacklustre response to Diplock. He wanted to clarify if 'trainers' could be classed as mercenaries, and whether KMS assistance to the Sri Lankan army would be perceived in the UK and, more importantly, India, as British government involvement. The department then gave Cleghorn the important task of consulting the government's legal advisors.[32]

Deep inside Whitehall, a process was now in motion whereby some civil servants and politicians would grow increasingly concerned about the conduct of KMS in Sri Lanka. Running parallel to this process, other Whitehall insiders would passionately advocate for KMS. Chief among the company's supporters was undoubtedly Sir William Harding. The paper trail, which the Foreign Office only declassified at my request, and then only partially, confirms that Harding did indeed speak to Colonel Johnson, again in mid-January. The KMS chairman was about to embark on a visit to Sri Lanka to discuss 'a sizeable commitment' in their training of both the police and army. The number of STF instructors would swell to 16, and a further five KMS men were now needed to 'retrain' an army commando unit – almost certainly the squad originally coached by the SAS after the Iranian embassy siege.

In a sign of the company's growing importance, Johnson said he had established a lieutenant colonel 'in an office adjoining that of the minister for national security, who had direct access to [the minister] Athulathmudali and was otherwise being "well looked after"'. The name of the KMS officer is censored, but as the most senior KMS employee and having retired from the SAS just three months before, we can deduce that it was probably Brian Baty. He was by now embedded with one of Sri Lanka's most powerful politicians, who played a leading role in the war effort. This was not enough for Colonel Johnson, who wanted to embroil his men even further. The company chairman hoped to recruit a senior quartermaster who could 'superintend the shipment of small arms and communications equipment' to Sri Lanka and bypass the country's current sloth-like process for purchasing weapons. In fact, the colonel was so keen to win the war that he was about to undertake a reconnaissance tour of Jaffna, possibly the most dangerous corner of the island, in order to better advise the Sri Lankans on what tactics they should deploy against the Tamil rebels. This maverick step did not appear to alarm Harding in the slightest. His only intervention was to remind the colonel of his promise that KMS personnel would not become involved in combat, and if he was considering diverging from that understanding then he must consult the Foreign Office first. Johnson readily agreed with this *laissez-faire* approach, and vowed to call the British envoy upon his arrival at the five-star Taj Hotel in Colombo. Harding remained deeply impressed by the former SAS commander. He signed off the telegram: 'Johnson is forthright, efficient and apparently very well informed about the situation on the ground in northern Sri Lanka.'[33]

Meanwhile the Foreign Office began to take stock of Britain's interests in Sri Lanka, to assess what was really at stake in this conflict. It forecast: 'ever-harsher repression of Tamil community in the north and further tension between Sri Lanka and India'. The latter consideration would turn out to be key. Although President Jayewardene supported an 'open economy' (i.e. capitalism) and followed 'a pro-Western line', the UK had relatively few commercial and investment interests in Sri Lanka. By contrast, Britain had 'substantial commercial and defence sales interests in India'. These interests, worth billions of pounds, could be jeopardised by 'Indian suspicion of any UK intervention' in Sri Lanka, because at that time India supported Tamil devolution. The Foreign Office was also determined to cover itself against any future liability, and feared that if it said nothing to KMS and allowed their training to expand to the army, it could 'lay us open to the charge that we were acquiescing in assisting the Sri Lankan army, the chief perpetrators of excesses against the Tamil minority'.[34]

And so, with these calculations in mind, and three weeks after Harding had given Colonel Johnson the green light to train Sri Lanka's army in their New Year's Eve phone call, the foreign secretary now wanted to row back. Howe met with Harding, Cleghorn and a handful of key officials, and decided to give the company 'a clear warning that Ministers were strongly of the view that KMS should not become involved in military training in Sri Lanka'. Howe also expressed reservations about Johnson's proposed reconnaissance visit to Jaffna. The strength of this warning might have been hampered slightly in that it was Harding, the KMS admirer, who was trusted with delivering it. In any case, the assembled diplomats realised the government had no legal power to stop KMS. Instead, the priority was merely 'to be able to make clear in Parliament if necessary that we had done all we could to dissuade the company from becoming involved in military training, given the likelihood that the military situation in Sri Lanka would continue to worsen and the real possibility of further killings of civilians by the Sri Lankan Army'. They also agreed to relay the contents of this meeting to Downing Street, indicating that the dilemma was shared with the prime minister herself.[35]

This crunch meeting took place the day before Johnson left for Sri Lanka, but no steps were taken to seize his passport – a discretionary legal power the government could have used if it was really determined to stop or at least disrupt KMS, but which would only ultimately be considered by the Foreign Office some 15 months later. Instead, Johnson touched down in Colombo and immediately got to work. It would be several more days before he deigned to contact the British high commissioner there, contrary to what he had promised Harding. He finally came to the ambassador's residence one evening for a heated *tête-à-tête*, which went on until midnight. It did not go well. Johnson was 'adamant' that his company should not restrict itself to only instructing the police, and insisted he had already committed to training the army in light of his New Year's Eve phone call with Harding which had given him initial approval. The envoy tried to explain that military training 'involved very considerable dangers', but Johnson was so intransigent that the envoy commented: 'I was talking to a brick wall.' The next day, the frustrated envoy met President Jayewardene and tried to convince him that recruiting more armed police, a *gendarmerie*, would be better than training new troops. However, the president already seemed convinced by the company's proposal to train thousands of counter-insurgency personnel. Now the envoy was furious. In stark contrast to Harding, he told London: 'Johnson is a glib, plausible and dishonest salesman.' The envoy speculated that Johnson had deliberately avoided him upon arrival in Sri Lanka, so that he could 'get to' the president first and seal his commitment to training the army. Among the embassy staff, it was not uncommon for the KMS directors Johnson and Walker to be referred to mockingly as 'Bodie and Doyle', after maverick characters in the popular TV crime drama *The Professionals*.

The envoy was scathing about Johnson: 'It is quite clear to me that despite his service as a Guards officer and aide-de-camp to the Queen, British interests carry no weight whatsoever in his manoeuvring for more and better business for KMS.' He grumbled that Johnson had persuaded the Sri Lankan military not to purchase the British army's troublesome new rifle, the Enfield SA80, warning them (quite truthfully) that it was 'rubbish'. Instead Johnson urged them to buy the more reliable US-made M-16, 'sales of which he could arrange'.[36] The only silver lining was when Johnson promised to see the British ambassador and defence attaché in ten days' time, before he left the island.[37] The ambassador need not have been so concerned. Johnson came back to him just two days later, appearing 'a little apprehensive of what, if any, sanctions we might take against him for his intransigence'. The KMS chairman had undergone a slight change of heart, and agreed with President Jayewardene that there should be 'regular discreet meetings' between the British embassy's defence attaché and the KMS 'resident', Brian Baty, who used the '*nom-de-guerre* of Ken Whyte'. These meetings would 'keep us completely up-do-date on what KMS are doing', although the envoy added sceptically: 'it will be interesting to see to what extent the information they pass on to us is as complete and frank as he indicated'.[38]

Meanwhile, back in Whitehall, the Foreign Office's lawyers reassured colleagues that Britain was not bound by any international treaties regarding mercenaries. Conveniently, although an ad hoc UN committee was trying to draft a convention on mercenaries, which 'might in certain circumstances cover the type of activities which KMS are getting involved in', the text would take two years to finalise and the UK was involved in the drafting process. The lawyer concluded that the international obligations the UK could use to stop KMS were 'virtually none'.[39] Cleghorn, clearly relieved, said: 'the problem would now appear to be one of presentation: how would we demonstrate publicly that we had sought to discourage the firm if their activities attract criticism in the UK?' Thus far, the Foreign Office had only given Colonel Johnson oral warnings. Was it now time to send the company a written warning?[40]

At this point, Britain's most senior diplomat hastily intervened. Sir Antony Acland, who had remained largely silent since KMS had won the contract to train the STF almost 18 months previously, scrawled on the back of a telegram: 'I do not think that writing will add anything … we have said enough to them urging caution and restraint.' Besides, there was a much more civilised way to deal with KMS. Colonel Johnson's best man, Sir Anthony Royle, had gone on to become vice-chair of the Conservative Party from 1979 to 1983, before moving into the private sector as a director of Westland Helicopters. He was now a peer of the realm, known as Lord Fanshawe, and would call on Acland about other matters. Such an encounter provided the Foreign Office chief with the perfect opportunity to ask him about Johnson. Royle remained loyal to his best man, saying Johnson: 'always tried to act correctly and to be sensitive to difficult political situations'. However, the peer took the Foreign Office's concerns on board and 'readily undertook to have a word with Colonel Johnson'. The old boy network had swung into action.[41]

Nonetheless, Britain's envoy in Colombo remained unconvinced. Johnson was a 'tricky and dishonest individual', he protested, 'and I certainly do not believe anything at all that he says'. Even Sri Lanka's director of national intelligence had told the embassy he was 'not too impressed' with the performance of KMS and fretted that one of the men may 'by accident or design' end up in a situation where he 'either gets shot or has to do some shooting himself'. The envoy instructed his defence attaché to have 'discreet and *irregular*' meetings with Brian Baty, and make it clear that ministers disapproved of the company's expanding role. The envoy ultimately concluded: 'I do believe that it is greatly to our benefit to know what KMS are up to here so that we can warn you of any developments that could further embarrass HMG. You may be assured that every precaution will be taken to avoid any possibility of the appearance of collusion or approval on our part'.[42]

However, this telegram was given a deliberately narrow circulation, marked 'secret and personal' and only declassified 34 years after it was written. The careful censorship, and destruction of associated files, suggests we cannot be certain this is the full picture of the British state's position towards KMS at that juncture. The company's chairman, Colonel Johnson, had worked extensively with the SAS and MI6 at various points of his career. Both organisations have considerable influence on British foreign policy with MI6 specialising in 'deniable operations', and yet neither is required to release their files to the National Archives. Accordingly, one must treat the available paper trail with a healthy degree of scepticism. Indeed, when I interviewed Lieutenant Colonel Richard Holworthy, who was

Britain's defence attaché in Sri Lanka from 1985 to 1987, he gave a rather different version of the arrangement between KMS and the Foreign Office. Speaking frankly and without the inhibitions of Whitehall censors, he said of KMS: 'They provided training teams and assistance in a lot of places, not always with the full cognisance of the British government. In our case the people who were organising it came out to the High Commission and told the High Commission what they wanted to do and they were allowed in.' He then added: 'They'd probably have come in if they weren't allowed in, but they were allowed in – so it wasn't official but they were there and they were doing a good job.'[43] From Holworthy's perspective, there was a greater degree of official toleration for KMS than the sanitised paper trail at the archives would have us believe. Perhaps it is this subtle but crucial distinction that Whitehall's censors have worked so hard to keep secret. Unfortunately for them, Holworthy had plenty more to tell me.

6

The Exploding Hospital

The Contras' fightback came on 6 March 1985. At quarter to eleven that night, Nicaragua's lakeside capital Managua was shaken by a series of explosions, the largest it had seen during the civil war. The blast centred on the capital's main military complex that housed the army's headquarters, a munitions dump and one of the city's main medical centres. The Alejandro Dávila Bolaños hospital cared for both military and civilian patients. A doctor on duty there that night, Sergio Martinez, told journalists from *World in Action* that he had to evacuate 150 patients, by guiding them along the hospital walls. His actions gave them some protection when the largest explosion threw 'bricks and huge lumps of rocks all over the place'. Although three of his colleagues were hit by glass from blown out windows, no one was killed.[1] Another witness, Charles Castaldi, told the TV programme:

> We could see that the whole hillside was engulfed in flames, 15–20 foot flames. It was quite chaotic around here. And as I started talking to the soldiers, there was a tremendous explosion. It was so strong you couldn't really hear it, you could feel the vibration, you could feel this impact, the impact of this wave. Everybody started running. I think the soldier that I was talking to threw his helmet off and started running down the street. Blocks of concrete, huge blocks of concrete started falling all around me. Like I even remember one hitting the street next to me, leaving a huge gash in the pavement. It seemed like the end of the world.

The blaze spread to the surrounding hillside, and could be seen from all over the capital.[2] The authorities counted at least three explosions, 'the first of which blew out windows half a mile away', and subsequent blasts were even more powerful. The explosions toppled power lines, cutting electricity for three miles. Debris was scattered 500 yards from the crater, which measured ten yards in diameter. A local resident reported seeing a fireball 200 feet high.[3]

The Nicaraguan government responded by downplaying the incident, not wishing to add weight to fears that the Contras had infiltrated the capital. They set up an inquiry that blamed faulty wiring. Later that year, after the incident faded from the headlines, their embassy in London issued a press release saying that a state of emergency had been decreed. One of the four reasons given for this measure was to 'neutralise the orders given by the United States to mercenaries in the sense of carrying out terrorist attacks "if possible near the capital Managua"'.[4] As it turned out, this press release was remarkably accurate. When the US Congress eventually began looking into Colonel North's covert actions in Nicaragua, they heard testimony from one American citizen who worked as North's trusted emissary of secret messages to the Contra leadership. This tall, athletic man, codenamed 'TC' or 'The Courier', was Robert Owen – a hawkish, ideologically zealous 33-year-old anti-communist from Rhode Island. A graduate of Stanford University, he worked as a private school tutor before going into politics as an aide for a powerful Republican senator. This job brought him into contact with Oliver North, while lobbying on behalf of Nicaraguan refugees in his senator's constituency.

Owen was a man of strong views, unrepentant and staunchly supportive of Colonel North. 'The totalitarian dictatorship in communist Nicaragua is a strategic threat to the rest of Central and Latin America and the United States. Make no mistake, the hardcore Sandinistas mean our nation harm', he lectured Congress earnestly, wearing a red tie and blue shirt, with his wife watching him from the gallery.[5] 'Theirs is a government that rules by torture and terror, fear and force.' The fact that Owen saw nothing wrong in what his colleagues had done in Nicaragua meant that he gave frank and full testimony, even on matters that many would find abhorrent. Among his revelations was that Colonel North had 'made a suggestion that his British friend and some of his people had been involved in what I guess you would call sabotage work in Managua'. In a closed session, that was not televised, the courier elaborated further. 'At that time it had been announced in the newspapers that there had been several explosions in downtown Managua and the Sandinistas were trying to say it was near a hospital and they were just minor explosions', Owen recalled. 'But then he [Oliver North] mentioned that some of his friends had caused them, I believe it was an ammunition dump to be blown up.'[6] This testimony would seem to implicate North's 'British friend' David Walker in the hospital blast.

Owen's extraordinary claim fits with a secret memo which North had penned just two weeks after the explosion, in which he outlined a slew of actions that were aimed to influence a vote the US Congress was soon to take on whether or not to restart funding for the Contras. Among the actions North listed were 'special operations attacks against highly visible military targets in Nicaragua'.[7] This description would seem to encompass the attack on the army headquarters adjacent to the capital's hospital, and would fit with the 'special operations' that North was eager for David Walker to be hired to carry out months before the bombing.

Although the Sandinista authorities knew that the fire at Managua hospital was never due to faulty wiring, the revelation of British mercenary involvement deeply angered their anti-imperialist sensitivities. Alejandro Bendana, general secretary of Nicaragua's Foreign Ministry, told *World in Action*:

I am surprised at how far the arm of the Reagan administration can reach, and reach out to the scum of this earth. Be it Colombian drug dealers or British mercenaries. With or without the approval of the governments, although those governments must reflect on their capacity to uphold their position as leaders in the struggle against terrorism when they in fact allow terrorists to be recruited in their own territory.

With a flourish of rhetorical mischief, he asked: 'what would the British government think if we allowed the most recalcitrant elements of the IRA to have a nice little base here in Nicaragua?' Bendana continued: 'We are absolutely shocked that an Englishman could have stooped to this and that evidently he has not been brought before justice. We think that there is an account that has to be paid here.' The foreign minister said David Walker had no 'morality at all' in his dealings with Nicaragua and that British people should see David Walker 'For what he is, a terrorist'.[8]

Whatever the Sandinistas thought of him, Walker was only just beginning his company's dirty war in Nicaragua. He had clearly been busy in Central America for some time, and this once diligent local politician had not shown up to a council meeting in the six months before the hospital bombing. Remarkably though, a week after the explosion in Managua, Councillor Walker was back at Elmbridge Town Hall. He attended a Recreations and Amenities Committee meeting, where they agreed to support a campaign to save Surrey's swans, who were being poisoned by lead fishing weights. They also discussed the issue of fire safety at Weybridge library.[9] By now, Walker was leading a double life. At home, he was a committed conservationist. Meanwhile, abroad he was allegedly setting civilian installations ablaze, and showed no sign of ceasing his commitment to the Contras.

Two days after Managua's hospital exploded, an arms manufacturer in Belfast, Short Bros Ltd, applied to the UK government for permission to sell Blowpipe surface-to-air missiles to Honduras – Nicaragua's northern neighbour, which was a key US ally in Central America and the main Contra base in the region. The application took several months to progress through

Whitehall's bureaucracy, eventually landing on the foreign secretary's desk. Although his colleagues in defence procurement and the Northern Ireland Office saw the potential for job creation, Geoffrey Howe was more concerned about the diplomatic ramifications of such a deal. Despite Thatcher's close relationship with Reagan and her fear of Nicaraguan leftism, by 22 July 1985 Howe had reached the conclusion that he was unwilling to agree the sale of Blowpipes to Honduras because there was a risk of the missiles 'falling into the hands of the Contras'. The UK had publicly committed to opposing an arms race in Central America, and one diplomat pointed out how the Contras were using Honduras as a major supply route.[10] Walker's Central American business plan was running into some obstacles, but his commitment to the Contras would not end there.

* * *

Batticaloa, October 1984

At first they were welcomed. Local people thought the new police unit was 'potentially an improvement on the army presence'.[11] They did wear a strange uniform though. A special camouflage print, different to the plain khaki of most police. Their weapons looked far more modern too. What a novel sight they were as they patrolled the streets of Batticaloa in their olive Land Rover Defenders. Batticaloa was not as war-torn as Jaffna, and was better known for its water rather than pools of blood. The Eastern Province's flat landscape was pitted with shimmering lagoons, from which locals said they could hear the fish sing. The legendary acoustic phenomenon was so widely accepted that a roundabout in Batticaloa's town centre was ringed with stone fish, bellowing their lungs out beneath a squat clock tower.

It was not long before the STF began to make their presence heard above the mythical chorus of the fish. 'Police Commandos have arrested a hundred suspects in the Batticaloa area', a British diplomatic telegram fizzed.[12] The local militant groups began to fight back, staging night-time ambushes.[13] There was a plethora of political thought among the Tamil revolutionaries in the East, and it was one of the more socialist groupings, the Eelam People's Revolutionary Liberation Front (EPRLF), which was one of the first targets of the STF in Batticaloa, with four of its cadres including a local leader soon slain.[14] On the other side of the island, four British diplomats at the embassy in Colombo were carefully monitoring this escalating violence: John Stewart, the high commissioner; Justin Nason, his deputy; Richard Holworthy, the defence attaché; and Alasdair Tormod MacDermott, whose job description was more ambiguous. By December 1984, one of the team noted that complaints about mass arrests had become 'more frequent with the STF itself being specifically criticised'.

However, Merril Gunaratne, who as head of Sri Lanka's intelligence apparatus in 1985 effectively oversaw the STF, did not show any such concerns about this period when he spoke to my team: 'They were meeting with success, I remember in the East.' The commandos worked in small groups to catch the Tamil rebels off guard, a daring tactic Gunaratne glowingly attributed to their training by the British mercenaries. 'The investment with the KMS was profit worthy', he insisted. 'We could never have reached the standards we reached, we could never have generated a fear psychosis in the LTTE with the STF if they were locally trained. It was the foreign training, the KMS training, which gave them the expertise, which gave them the confidence "we are second to none".'

This sudden surge in the security forces' potency certainly had an acute effect on the rebels' morale in Batticaloa. The PLOTE, a more Marxist offshoot of the LTTE, rallied its followers to resist the onslaught. 'My wish and message to all the people of Tamil Eelam is that we will not allow the temporary setbacks and government oppression to weaken our resolve in the struggle which lies ahead', marshalled comrade Uma Maheswaran, who had originally founded the LTTE with Prabhakaran before breaking away. In his New Year's message, the comrade counselled: 'I will not like a few bourgeoise [*sic*] leaders misinform you about the date and time and the day of liberation. Our party is not surprised at the actions of the Sri Lanka government. It is the natural sequence of events.' His prognosis was both bleak and hopeful. 'In the year ahead lie blood, sweat and tears. This is our immediate future. But let us not forget, and let us not lose heart, for at the end of [the] tunnel is the light of day and freedom … Liberation is at hand. Victory will be ours.'[15]

Maheswaran was right about the blood and tears, but not much else. Allegations of murder by the STF soon began to emerge, and by March 1985 the Foreign Office was sufficiently curious to dispatch one of their team to take a closer look. MacDermott, 39, was a dashing Scotsman with a shock of brown hair, a neatly trimmed beard and de rigueur sunglasses. He also had a flair for languages, and had studied Japanese at the School of Oriental and African Studies, a bohemian campus in central London. His career in the Foreign Office was rather more conservative however, and some people who knew him in Colombo say he was working for MI6. His reportage after visiting the East was titled ominously 'Batticaloa – the next Jaffna'. In it, he warned that, 'The situation in the Eastern Province resembles that of Jaffna a year ago. It has deteriorated markedly in the last six to nine months'. MacDermott singled out the STF to blame for this state of affairs, and said they gave him 'most cause for worry'. The police commandos had almost taken over the region, leaving the civil administration 'emasculated'. Compared to the army, the STF was now 'regarded as if anything worse and is feared and disliked'.

They had lost the initial good will of eastern folk once it emerged that their tactics were 'strikingly similar' to those used by the army in the north, including mass arrests of all young Tamil and Muslim men. MacDermott attached reams of allegations against the STF that had been carefully recorded by the local citizens committee, a panel of professionals from doctors to lawyers and clergy. It seemed as though sending an all-Sinhalese force to a predominately Tamil-speaking part of the country had not been Colombo's wisest move. 'The STF are generally believed to regard the population of the Eastern Province as hostile, the area as foreign territory and the youthful population as potential terrorist suspects, one and all', MacDermott explained. However, it would be hard to challenge the STF's conduct, because they were 'believed to be under the direct orders of the Minister of National Security. Appeals to him for restraint in the use of the STF have been met with the response that their methods are necessary to combat terrorism.'

As much as MacDermott wanted to paint this as a Sri Lankan problem, he was aware of the potential for unwelcome ramifications in London. 'The extent of local dislike of the STF is such that any signs of HMG supporting its activities or its role would be deeply resented at all levels in the Eastern Province and would risk losing us any residual sympathy or influence we have with the populace of the area', he warned, in a veiled reference to KMS. 'The appearance given by the STF of being an army of occupation has effectively blinded the judgement of those who would normally be critical of violence. Not even the church will condemn the terrorists outright but concentrates its criticism on STF excesses.' The unit's behaviour, it seemed, had earned it 'fear and loathing in a mere six months'. If this was not bad enough, the Scotsman went on to make a damning indictment of KMS training: 'It is in fact the STF which is the most effective recruiting agent for the terrorists today', in spite of them receiving superior training than the army.

To make sure his telegram was noticed, he heavily hinted that KMS was not helping matters. 'Rumours of training by ex-members of the SAS led to hostile questions from several people representing the more moderate organisations of the area.'[16] Tellingly, I have found two copies of MacDermott's report at the National Archives, and in one copy that remark remains censored. Clearly, what he had seen on his trip to Batticaloa in early 1985 was enough to cause consternation among his usually restrained colleagues. Bruce Cleghorn, the civil servant turned censor, was involved in this discussion at the time, and described it as 'a very worrying, but not unexpected, development'. The deputy high commissioner, Justin Nason, commented: 'We are particularly concerned about the accusations against the STF and the fact that the STF is believed by the people of Batticaloa to have been trained by ex-members of the SAS now working for KMS.'

However, Nason was creative in coming up with possible explanations for what had gone on. 'My own view, for what it is worth, is that the STF may not be as bad as the Tamils of Batticaloa make out', he reasoned. Perhaps they were just following orders issued by Sri Lanka's national security minister, 'to arrest and detain all young men between the ages of 18 and 30'. If this was the case, then he believed that the STF would not hold them for long before handing them over to an army camp at Boosa, 150 miles away to the south. As such, he felt the lack of information available to relatives about the location of their young men was 'neither the fault nor the responsibility of the STF'. Despite this inventive theory, the deputy envoy pledged to keep an open mind, admitting: 'I may well be proved wrong when we receive the documents promised by the Citizens' Committee.'[17] With or without the affidavits, it is a telling indication of the Foreign Office's attitude towards human rights that this senior British diplomat did not regard a blanket order to arrest all young men, effectively internment, as maltreatment in and of itself.

MacDermott duly forwarded the testimony from the Batticaloa District Citizens Committee, commenting: 'The change in attitude to the STF from one of public confidence to one of almost universal dislike is disquieting.' However, he also added a rather large caveat: 'We are not sure how much credence to give to claims of ill treatment made about the STF (as opposed to the army and prison officers). We must, of course, guard against becoming ourselves victims of Tamil propaganda.'[18] The evidence compiled by the citizens committee, which ran to over 50 pages, appeared lost on the sceptical diplomats. The committee, which mostly comprised Catholic priests, had spent hours sitting with distraught Tamil families, recording in painstaking detail scores of arrests, disappearances, torture and in some cases deaths. Although the diplomats were unimpressed by it, this trove of paperwork paints a vivid picture of the harassment and hardship that was inflicted on the Tamils of the Eastern Province at the time. There were so many cases that it is hard to calculate with any confidence how many young Tamil men vanished at the hands of the STF back then, but one human rights lawyer I have spoken to in Sri Lanka had records of 300 disappearances from the three years the unit was being trained by KMS. That figure is likely to be an underestimate, as this was only the data that his small law firm had been privy to.

Other sources told me anonymously about a smaller number of cases they were aware of, but the intensity of the abuse magnified the scale of the suffering. A doctor who treated torture victims told me nonchalantly about ten cases of people who had been hung from their thumbs by the STF in those dark days. A form of waterboarding was also used, among cruder techniques such as beatings. Women were stripped naked, others were burned with cigarette butts. Another person had his penis put in a drawer and slammed shut. This abuse all happened behind closed steel doors and barbed wire fences, only to emerge much later upon release from custody. But many of the STF's onslaughts took place in plain sight and could not escape anyone's attention. One affidavit attests to a 'massive comb out operation' that took place in Komari, a coastal village south of Batticaloa, on 24 February 1985, in which 500 commandos and four helicopters scoured the area. Any youth found in their path was rounded up and made to 'shout slogans condemning the terrorists', in a crude spectacle of collective punishment.

To their credit, the British embassy team at least skimmed through these reams of testimony. The defence attaché, Lieutenant Colonel Richard Holworthy, noted: 'Interesting to hear the other side of the Komari combing out op – after report to me by KMS', suggesting that the British mercenaries had given him a different version of events. Among the dense dossier of muted horror, some stories stand out. A 15-year-old boy, Poopalapillai Ramanathan, was shot dead along with two friends when the STF opened fire on their wooden canoe while they were rowing in a lagoon near the village of Koddaikallar. Other youngsters were taken to a beach, where their faces were buried in the sand before they were shot. When their parents came to the hospital mortuary to identify them, they found sand stuck inside their children's mouths.[19] Another case is even more chilling. A young girl, Illayathamby Sashikala, was allegedly stripped naked and asked to lie on the ground as STF men stood on her legs with their boots. They lit a fire with coconut palms and carried her near it, 'threatening that she would be dumped into it'. The traumatised girl was then detained at Batticaloa police station for three weeks before being released. She was ten years old.[20]

One of the most forensic accounts available is a report by a ceasefire monitoring committee, which visited Batticaloa in November 1985 during a temporary truce. This lull in fighting allowed them to speak to both local people and STF constables. They examined in detail one incident where a land mine exploded on Lake Road, near the centre of Batticaloa Town, seriously injuring five STF men.[21] In echoes of the Point Pedro massacre a year before, the STF then opened fire and nine civilians were killed. A disturbing pattern of vengeance and cover-up was emerging. Although several of the STF men had been knocked out by the initial landmine blast, one remained conscious and managed to fire 29 bullets towards where the corpses were found. Other officers described giving covering fire as if it were a crossfire situation.

Their version of events was contradicted by a witness who says that after the landmine exploded he saw the STF force five people to walk down a road at gunpoint before executing them. Another survivor described how he managed to get his son to come inside their house merely seconds before an STF commando fired a bullet through their door. He later saw nine of his neighbours' bodies at the hospital. The dead Tamils from this incident included Manojkumar 'Maju' Pirathapan, a student at St Michael's College who was two weeks shy of his nineteenth birthday. The post-mortem reports for the nine men, aged between 16 and 30, found that six were shot from the back, and four were killed by a single bullet each. The committee said that the STF's version of coming under fire after the landmine exploded was 'extremely improbable' and noted that none of the other police jeeps were damaged in the supposed crossfire, concluding that the STF's evidence was 'far from convincing'.[22] Local people had no doubt who was responsible and around a week later, another land mine hit an STF patrol, killing several commandos. The ID cards of the dead Tamil civilians were found on their corpses.

These incidents clearly fed a cycle of violence that left a deep scar on survivors. Even those Tamils who had some protection from the STF, by virtue of their status as priests, embarked on the gruelling process of documenting in meticulous detail what was happening to their flock. Father John Joseph Mary is a diminutive figure simply dressed in a white cassock, but with a fearsome reputation for standing up for the rights of his people. Now in his mid-eighties, he was inspired as a young man by liberation theology, the brand of left-wing Christianity that was making its mark on peoples' struggles across Latin America. The movement forced the Vatican old guard to acknowledge the Church's duty to challenge oppression. He now lives in a sprawling Jesuit compound on the outskirts of Batticaloa, overgrown with vegetation. It can house hundreds of church workers for educational events and spiritual gatherings, but was eerily quiet when I arrived, the sheet metal gate swinging open without anyone touching it. In the sweltering heat of a Sri Lankan summer, the veteran priest could talk for hours about all aspects of the conflict, although we finally focused on the history of the STF in Batticaloa in the 1980s.

'There's a record of missing people', he sighed. 'From deep villages they would come: "my husband has disappeared, my son has disappeared". By whom? "By STF". People have been killed. They are just ordinary poor people who earn their living hand to mouth.' Sitting forward in his chair, he calmly recounted a harrowing hallmark of the STF that had been sanitised from the British files. The commandos would place tyres over their prisoners before setting fire to the rubber wheel, until the body was 'burned completely', he whispered. 'Completely. So, that's what happened. Everywhere: tyre, tyre.' The victims of these gruesome executions were not militants. Father Joseph Mary insisted: 'They didn't belong to LTTE or anything. Ordinary citizens.'

The barbaric practice was known as 'Necklacing', and it was fast becoming the STF's macabre calling card across the Eastern Province. Although there is no suggestion that KMS taught or encouraged the STF men it trained to execute prisoners in this manner, the practice seems to have persisted in the years that the company trained the commandos, as much as the British embassy staff had tried to dismiss the allegations of abuse as 'Tamil propaganda'. When I asked Father Joseph Mary what he made of that assessment, he was visibly shaken. 'That's the first time I've heard that', his voice trembled. 'That's shocking, shocking.' He turned the tables and suggested it was the British diplomats who were producing the propaganda, on behalf of the STF, to gloss over the impact of the mercenaries. 'It's we who have suffered who know', he concluded witheringly. 'Those who are sitting in their office under a fan in the ceiling just taking a report and making a comment – they can make any comment.'[23]

*   *   *

Supposing British diplomats in Colombo had sincerely wished to raise concerns about the activities of the STF and its KMS instructors, the opportune moment would have been in March 1985 when one of the most senior figures in the Foreign Office, Sir William Harding, was preparing to visit Sri Lanka. In a telegram that is remarkable for its frankness, the veteran diplomat reveals that he lunched with the KMS 'managing director', Colonel Johnson, in London.[24] They were joined by Major David Walker – a mere nine days after the hospital bombing in Nicaragua.

The men appeared to get on well, and spent most of the start of their meal discussing the origins of KMS, although frustratingly Harding did not record the meaning of the company's mysterious name. It turned out that one of the company's co-founders, Brigadier Mike Wingate Gray, had previously served alongside Harding as a military attaché in Paris, a crossing of paths that would endear Harding to the company more than he already was. With the pleasantries and name dropping out of the way, Johnson began to express President Jayewardene's 'keen disappointment' at the British government's refusal to 'come to his aid in his present travails'. However, Harding was not persuaded to change course, and explained that Britain's relations with India overshadowed the whole affair, because the UK had interests there that were of 'immeasurably greater magnitude than those in Sri Lanka'. At that moment, Whitehall was trying to secure billions of pounds worth of arms and equipment sales to India. Harding lectured the KMS men that the Foreign Office's 'best efforts must be directed to getting off on the right foot with the new Indian government and Rajiv Gandhi', a world leader who did not welcome the presence of British mercenaries in his back yard.

The conversation then turned to what KMS was actually doing in Sri Lanka. It is here that the censors start to redact the telegram. For instance, it states: 'Colonel Johnson said that Whyte … [four lines redacted] was acting effectively as the Director of Military Operations of the Sri Lankan High Command.' Notwithstanding the redaction, this appears to suggest that Whyte (i.e. Brian Baty) was by now running the war against Tamil rebels. Assuming Colonel Johnson was not exaggerating, and Baty did in fact hold such a senior position as director of military operations, it would implicate Baty with command responsibility for war crimes that Sri Lankan forces were perpetrating in March 1985, such as the disappearance of Tamil detainees in the east. The telegram goes on to reveal that KMS had hired a former British army intelligence corps officer and ex-Gurkha, Lieutenant Colonel Joe Forbes, who was 'acting as Director of Military Intelligence', another incredibly senior role in the war effort, redolent of responsibility for the atrocities that were taking place (and about to worsen). However, the company's influence did not stop with these two prominent postings. It emerged that Colonel Johnson had attended a cabinet meeting of the Sri Lankan government where, at the invitation of President Jayewardene, he explained 'exactly what, in his view, was wrong with the Sri Lankan armed forces'. Johnson also claimed to have 'very open and friendly relations' with the president's son, Ravi, who oversaw the STF, as well as maintaining 'very close' relations with the powerful national security minister, Lalith Athulathmudali.

Harding was impressed: 'KMS seemed to have been very successful in establishing a confidential relationship with the senior power structure of the present Sri Lankan regime.' The company was so deeply embedded that Harding warned them about the risk of being targeted by the opposition if their positions became too high profile. The company now had dozens of staff in Sri Lanka, mostly coaching the police and the STF, although some were retraining the army commando unit. By the end of the meeting, Harding seemed enamoured with the KMS leadership. 'I took to Colonel Johnson rather more than [the British High Commissioner to Colombo] Mr Stewart appears to have done.' In a remarkable comment, Harding said that although Johnson's 'political ideas are probably to the right of Ghengis [*sic*] Khan', he did not think the rapacious mercenary was 'trying to hoodwink' him. Harding signed off: 'It is much better to get in touch with KMS and wield whatever little direct influence we may have on their activities, rather than keep them at arms length.'

Shortly after this intimate luncheon with the KMS leadership, Harding visited Sri Lanka himself. Once in Colombo, he met Britain's defence attaché, Lieutenant Colonel Holworthy, who had read the reams of complaints about the STF in Batticaloa. Despite knowing the allegations against the company, Holworthy did not sound the alarm with his superior. Instead, Holworthy 'spoke highly of the limited work being carried out by KMS Ltd'.[25] This was a crucial missed opportunity to reign in the company, but none of the Foreign Office's Sri Lanka team bothered to speak out. The high commissioner himself even advised Harding not to talk to the company during his visit, effectively avoiding the issue altogether. In the end, Harding seemed so at ease with the company's work in Sri Lanka that he concluded, in a manner reminiscent of the Diplock debate: 'We can leave the KMS to get on with the training on the spot which they appear to be conducting to the great satisfaction of the Sri Lankan authorities.' If he was aware of the complaints about the STF, he must have thought they were propaganda, because he said: 'As the Sri Lankans seem singularly inept at putting their own public relations case, we might also consider whether we can offer them some discreet training in information techniques.' Although it would be easy to accuse the British High Commission staff of inaction, it is highly unlikely that Harding would have listened to any of their concerns about the company. The mandarin was intensely ideological and deeply resented the Tamil liberation movement, largely because he believed it was Marxist.[26] When Harding met Sri Lanka's defence secretary, General Attygalle, he stressed the need for 'a proper campaign of psychological warfare exposing the Marxist leanings of the terrorists'.[27]

Upon his return to London, Harding spoke to Colonel Johnson by telephone, who boasted that the commandos his company had trained were now 'operating effectively'. However, he warned that

India wanted President Jayewardene to turn away from the West and seek support from the Soviet bloc, a prospect guaranteed to alarm a Cold War warrior like Harding.[28] The Foreign Office knew that India resented actors from outside the region, such as KMS, interfering in Sri Lanka, because New Delhi felt the island lay 'squarely in their sphere of influence'.[29] Unfortunately, from Whitehall's perspective, if KMS pulled out altogether this could create a dangerous vacuum, which might be filled by China or the Soviet Union.[30] As Holworthy later explained to me: 'Quite frankly it was better that a British team went in to fill it [the training role] instead of somebody else.'

* * *

Divide and Rule

O mankind, indeed We have created you from male and female and made you peoples and tribes that you may know one another.

Quran, 49:13

In all conflicts there are tipping points. Thresholds that are crossed. Moments of not looking back. After April 1985, something profound changed. The Eastern Province of Sri Lanka is a remarkably diverse area. There are of course the Tamils, who are Hindu or Christian, and a smaller number of Sinhalese Buddhists. But there is also a third group. They speak Tamil, but identify primarily as Muslim. They occupied a middle ground and could sometimes escape the Tamil–Sinhalese clashes. In many parts of the Eastern Province, the Muslims enjoyed intimate interactions with the Tamils, in some cases going as far as intermarriage or undergoing religious conversion. For example, the town of Eravur, which neighbours Batticaloa to the north, is said to have evolved after an elder had a vision of Muslim–Tamil unity. According to local folklore, the elder is said

1. A poster produced by the Popular Front for the Liberation of Oman and the Arabian Gulf in the early 1970s

Credit: PFLOAG

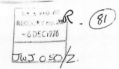

Mr M E Heath WAD

DIPLOCK COMMITTEE ON MERCENARIES

1.   Thank you for sending me a copy of your submission to Mr Mansfield of 29 November. I should be happy to join in a discussion on your Paper. May I make two points in advance of the meeting you propose to call?

2.   First, I should like to repeat that our firm preference is still for no legislation at all. As far as my Department's area of interest is concerned, legislation would provide no benefits and might attract substantial disadvantages. It is difficult to quantify the latter, but my feeling is that in exchange for mitigation of the type of publicity which caused us embarrassment over Angola, we might end up with a situation which could do real damage to our bilateral relations with countries of political and economic importance to us. I accept (para 5 of your Paper) that Ministers may feel they have to do something. Nevertheless, I believe we should present to them the benefits of doing nothing.

3.   Secondly, there is the question of the private recruitment agencies which have come to our notice recently. One, KMS, is known to us, Security Department and PUSD, and has substantial business with a number of Middle East Governments. We have not been directly involved in the expansion of their activities and see no reason why official advice on the Diplock Report should be tempered to meet their requirements. However, their activities are a factor in the situation and I am therefore copying your paper to PUSD and Security Department, in case they wish to comment. As I see it, both options suggested in para 11 of your Paper could result in a situation in which HMG was required to approve or disapprove of the activities of agencies such as KMS; that might be a very difficult decision which could cause embarrassment both to HMG and to the Arab Governments for whom KMS is acting.

4.   I hope you will not mind if I also copy these papers to Mr Urwick, since NENAD have hitherto shared our objections to new legislation on the subject.

I T M Lucas
Middle East Department

2 December 1976

cc:
Mr Reid CSAD
Mr Laver, Rhodesia Dept
Mr Freeland, Legal Advisers
Mr Urwick NENAD   }
Mr Edes PUSD   }   with copies of submission and paper
Mr Howells, Security Dept)

2.   Foreign Office telegram from 1976 warning about the impact of any anti-mercenary legislation on KMS

Credit: Phil Miller

3. Sultan Qaboos at his Bait Al Baraka Palace in Muscat, Oman in 2010

Credit: US Department of Defense, rights free

4. Sri Lankan President Junius Jayewardene lands at a US Air Force base in 1984. He was nicknamed 'Yankee Dick' because of his pro-Western policies

Credit: US National Archives, rights free

5. David and Cissy Walker at the Lewa Wildlife Conservancy debate held in a South Kensington jewellers in 2012

Credit: Desmond O'Neill Features

6. In 1978, investigative journalist Duncan Campbell obtained photos showing two of the KMS founders, Brigadier Mike Wingate Gray (left) and Colonel Jim Johnson (right) outside the company's first office at 11 Courtfield Mews. His article revealed that the company was using this sleepy residential property for running a mercenary business, and soon afterwards KMS moved to a more permanent location in South Kensington on Abingdon Road

Credit: Photos by Chris Davies and others. Duncan Campbell's article was published by Time Out and is available at www.duncancampbell.org/PDF/Soldiers%20of%20Fortune.pdf

7.   As in Dhufar, women were an integral part of the Tamil armed movement

Credit: LTTE

8.   Merril Gunaratne was in charge of Sri Lanka's intelligence apparatus when KMS worked in the country

Credit: Lou Macnamara/Yardstick Films

9. Colonel Oliver North speaking at a US military base in Iraq in 2007

Credit: Wikimedia

10. US President Ronald Reagan sits with UK Prime Minister Margaret Thatcher at Camp David on 22 December 1984, where they discussed the situation in Nicaragua

Credit: US National Archives

11.   An unidentified KMS instructor with STF recruits; note their
M-16 weapons

Credit: JDS Lanka

12.   Another unidentified KMS instructor with Special Task Force
recruits

Credit: JDS Lanka

13.   The STF chief instructors' board records two English names as the first occupants of this important role

Credit: Lou Macnamara/Yardstick Films

14.   Athula Dualagala was originally trained by KMS in 1985 and went on to become STF director of training by 2018

Credit: Lou Macnamara/Yardstick Films

15.   Joseph Rajaratnam was a maths teacher at Hartley College in Point Pedro when the STF burned down its library in 1984

Credit: Lou Macnamara/Yardstick Films

16.   Alasdair MacDermott was a diplomat at the British High Commission in Colombo who recorded allegations of atrocities by Sri Lankan forces linked to KMS

Credit: Yardstick Films

17.   Jesuit priest Father John Joseph Mary described atrocities by the STF in Batticaloa, eastern Sri Lanka

Credit: Lou Macnamara/Yardstick Films

18.   Lieutenant Colonel Richard Holworthy was defence attaché at the British High Commission in Sri Lanka from 1985 to 1987

Credit: Lou Macnamara and Angus Frost/Yardstick Films

19.   The author with KMS veteran Robin Horsfall on the roof of his
flat in Prague

Credit: Lou Macnamara/Yardstick Films

20.   Anthony Knowles after his
arrest

Credit: Police handout

21.   KMS helicopter pilot Tim Smith

22.   A Sri Lankan Air Force Bell 212 carries a national flag at an Independence Day parade in 2019 while a door gunner keeps watch

23.   A memorial lists the names of all those who died at the Kokkadi-cholai prawn farm massacre of 1987

Credit: Lou Macnamara/Yardstick Films

24.   Former British high commissioner to Sri Lanka David Gladstone

Credit: Lou Macnamara and Angus Frost/Yardstick Films

25. The KMS/Saladin office on Abingdon Road in London
Credit: Phil Miller

30 August 2017

DIG Mr M.R.Latiff
Commandant, the Special Task F.
Sri Lanka

Dear DIG Latiff,

I am told that you will be holding a ceremony on Friday to commemorate the father of the STF Ravi Jayawardene, the first Commandant Zerni Wijesuria, Commandant Lionel Karunasena and DIG Upali Sahabandu. They were great men and good friends who played a major role in winning the war in Sri Lanka, and I would not want the occasion to pass by without paying my own tribute to them.

You may recall that my partner Colonel Jim Johnson and I were privileged to be able to make a contribution to the war effort. We arrived in 1983 on the invitation of President JR and remained throughout his time in office. Our main focus was in assisting and advising on the formation of the STF, though we were also involved in other areas. It continues to be a matter of great pride that I frequently heard STF described as the most effective force in the country. (At the time we operated as KMS Ltd - Saladin Security is its successor company).

We worked closely with Ravi, Zerni, and Lionel and I believe to an extent with Upali, though the latter achieved his promotions after our departure. The creation of STF was of course their achievement, and we were proud to assist. They were fine men and they made great sacrifices for your country. Ravi was a man of great, quiet force who was critical to the formation of the unit. Zerni was a brilliant first Commandant; wise, imaginative and devoted to the task. Lionel was headstrong, hugely determined and highly energetic, to the extent that he drove himself to an early death. I have no doubt that Upali was a fitting successor. I think of them often, and I admire them all.

I shall be thinking of you all on Friday. May God bless those fine men

With every good wish,

David Walker.

26.   Saladin letter to STF Commandant Latiff

Credit: Yardstick Films

27.   STF in riot gear advance on protesters, June 2017
Credit: Yardstick Films

28.   Grieving relatives hold a portrait of Sathasivam Madisam who
drowned while running away from the STF in 2017
Credit: Lou Macnamara/Yardstick Films

to have foreseen a shrine for a Tamil Hindu deity being erected at some lime bushes where a Muslim Sufi shrine would be built alongside. This dream led to the two places of worship being assembled and the settlement of Eravur growing up around them.[31]

By 1984, this mystic legacy was taking on new forms, with Muslim youth increasingly showing sympathy for the Tamil rebels, and the British files make frequent references to Muslims joining the LTTE or other militant groups. PLOTE was particularly optimistic about Muslims taking part in an island-wide proletariat revolution. 'In the eastern province, where for generations the Muslim community lived in conditions of want and poverty, a new leadership began to emerge', Comrade Maheswaran noted approvingly at the start of 1985. 'This leadership has grown out of the oppression and needs of the poorer sections of the Muslim community who in the majority are peasants and agricultural workers who are living on the fringes of society.'[32] PLOTE had received training in Lebanon with Palestinian factions – even its name was inspired by the Palestine Liberation Organisation – a solidarity initiative that could only endear it to the Muslims in the east of Sri Lanka. Comrade Maheswaran tasked one of PLOTE's most astute military analysts, Dharmeratnam Sivaram, with running strategy classes for PLOTE cadres, in which he stressed the importance of not alienating the significant Muslim minority in the east.[33] The LTTE were also well aware of the risk of division, and published an open letter to their 'Islamic brother' asking the Muslim community to 'beware of the Sri Lankan Government's attempt to drive a wedge between the two brother communities.'[34]

This potential for cross-community unity was not lost on the STF, who saw it as a strategic threat. And in April 1985, any prospect of Muslim–Tamil solidarity was shattered. 'I met six Tamil businessmen who had been burned out of their home on 23 April. They had lost everything', Britain's defence attaché Lieutenant Colonel Holworthy reported after a visit to

Batticaloa.[35] He went on to deliver an insightful account of the carefully crafted chaos:

> At midnight on 23 April, during the curfew, about 1,000 Muslims had approached the houses, armed with shotguns, and completely destroyed them. I asked if any Security Forces were present. They hedged and said they were afraid to say any more. When pressed [one of the Tamil business-men] Lingaratnam said that at the back of the mob were two army Landrovers and an armoured car. They did not join in but made no effort to prevent the burning and looting that followed. Once the mob had dispersed the troops remained to inspect the damage. When Lingaratnam remonstrated with them he said 'They just laughed'. I asked why the Muslims did this. He replied that he thought the Muslims wanted the Tamils out of the area because they were 'put up to it'. They had previously lived in peace and would never have done such a thing without provocation and support. The Tamils all said they would never return to that particular area. The distrust and fear is now very deeply rooted.

In days of unprecedented Muslim-on-Tamil violence in April 1985, 2,000 Tamil homes were destroyed and around 8,000 Tamils became refugees in their own country. British diplomats noted: 'It seems that the police commandos (STF) aided the Muslims and in some cases gave them arms and ammunition. We have even heard reports that the STF incited Muslim youths into attacking Tamils.'[36] Damning telegrams continued to hammer home the state's collusion in the riots. Holworthy himself noted: 'We are getting increasing evidence from many reliable sources that the STF were actively helping the Muslims in their attacks on Tamils in Karaitivu and Akkaraipattu', two towns south of Batticaloa.[37]

Around 30 Tamils were killed in these clashes, and the economic damage was valued at £8 million. British newspapers reported the incident and Sri Lanka's minister of home affairs was quoted in the *Times* and the *Guardian* as saying 'Police Commandos had been partly responsible for the attacks by Muslim youths on Tamils'.[38] However, the internal British diplomatic cables from the incident are even more critical. Holworthy spoke of a 'mounting dossier of written evidence' compiled by the citizens' committees about excessive STF violence in the east.[39] There was a sense of alarm about the implications for KMS and by extension Whitehall. British embassy staff believed the company was trying to play down the severity of the incident when it spoke to the mercenaries. The diplomats put more trust in other sources which said there was 'overwhelming evidence to indicate that there is a *prima facie* case against the STF'.[40] The High Commission team told London: 'In certain instances we believe that the STF aided and abetted the Muslims but we are not sure whether the orders to do so were given by a local commander or by central government.' Their concerns registered faintly in Whitehall, with one colleague noting: 'It appeared that the STF had aided the Muslims and in some cases given them arms and ammunition.'[41] However their boss, Harding, showed his usual sympathies towards KMS and was reluctant to accept evidence that police units trained by the company had been involved in atrocities.

Shocking as it was, this was not just mindless violence. It served a Machiavellian agenda. The UK High Commission commented sagely that from Colombo's point of view the rioting had been a 'success'. The embassy explained: 'Any possibility that the Eastern Province Muslims would support the Tamil separatist groups is now even less likely than it was before.'[42] It does seem hard to believe that the STF's inaction was an accident. By this stage, the police commandos were a highly professional force that should have been capable of stopping communal violence if it had wanted to. Indeed, one of the first things KMS had taught the unit was riot control. Whatever the causes, the consequences of this Tamil–Muslim schism were profound. Holworthy visited Batticaloa and drove some 300 miles around the Eastern Province in July 1985.[43] During his trip around the east, he saw 'fear and distrust, caused by ethnic violence and security force complicity with the Sinhalese and Muslim factions'.[44] He could not have known it then, but this tension between them would later lead to intense violence between Tamils and their former friends, the Muslims. The LTTE would go on to expel Muslims from Jaffna, fearing they were informants. A mosque in the east was bombed during evening prayers. If this was a deliberate attempt at divide and rule, then it worked. One only has to look at the 2019 Easter bombing of the Zion Church in Batticaloa by local Islamic extremists to appreciate the terrible outworkings of this division, and how far the Eastern Province had departed from side-by-side shrines in Eravur. One has to wonder what role KMS played in stoking these initial communal tensions. David Walker had visited Sri Lanka in mid-April 1985, shortly before the violence broke out. And perhaps more pertinently, there were KMS men in charge of both directing military operations and military intelligence – two of the highest-ranking roles that could have given the go-ahead for such divide-and-rule strategies.

President Jayewardene was well aware of the allegations against his men in the east, and tried to distance himself from them. He expressed concerns to Britain's high commissioner about the ongoing atrocities committed by his forces, which he said were 'undisciplined and ill trained'. He noted that command responsibility was an issue, with many officers unwilling to punish their troops for excesses. In a revelatory comment, the president said: 'He had hoped that the KMS training would have been more successful than it has been.'[45] Neelan Thiruchelavam, a Tamil lawyer and MP from the TULF was more scathing. He told the *Observer* newspaper: 'We are deeply concerned that former SAS officers are responsible for training the brutal, ruthless and undisciplined police commandos.'[46] It was now over a year since the SAS 'map-reading error' veteran, Malcolm Rees, had become chief instructor at the STF training camp, and the unit he trained had developed a formidable reputation. From the island's northernmost tip at Point Pedro, and then spreading across the vast Eastern Province from Batticaloa to Amparai further south, it had left a trail of civilian casualties in its wake. Simultaneously, its grim toll was rising on the island's west coast, at the desolate Mannar peninsula which juts out into the Indian Ocean and is crowned by a squat white lighthouse. Tamil clerics there reported that the STF was involved in looting and rape, and the unit operated in plain clothes and unmarked cars. MacDermott considered them to be effectively an undercover unit who were 'not apparently under the control of any of the usual authorities'.[47]

Although some might argue that one year was not enough for KMS to instil sufficient discipline in the new commando unit, further training by the company did not seem to stop the massacres, which continued well into 1986. Half a dozen miles inland from what is now a popular surfing beach, between 40 and 60 civilians were reported dead at some paddy fields in Lahugala, Amparai district, on 19 February.[48] In the official version of events, the Sri Lankan government claimed the incident began when the STF received a tip off that a Tamil militant group, the Eelam Revolutionary Organisation of Students (EROS), was active in that town and had killed some Muslims. (EROS was allegedly founded by the father of M.I.A., the rap-star.) In a bid to frame the security forces, the EROS cadres had tried to burn the bodies – as human bonfires were by now the STF's calling card.

The Tamils had a very different version. The dead were innocent paddy farmers, slaughtered by the STF – and EROS had photographs and video footage of the victims. Remarkably, British diplomats found the Tamil version more credible, and believed the STF was responsible for the killings. 'Suspicious features are that all the casualties were dead and no wounded have been admitted to hospital in the area', they reasoned.[49] This critical analysis of the incident only solidified with time, with a diplomat later noting: 'It is becoming more certain that the Special Task Force were responsible for the massacre.'[50]

*   *   *

Tucked away in a sleepy converted barn on the outskirts of Bordeaux, amid rolling French vineyards, I found Lieutenant Colonel Richard Holworthy, OBE. The OBE was awarded after his tour of Sri Lanka, a post he held from 1985 to 1987. Now in his early eighties, he is an active member of the local British Legion branch and enjoying his retirement. He was born in India during the British Raj, the son of a major general. The young Holworthy wanted to join the navy, until realising he was colour blind and would not be able to distinguish between the green and red bulbs of lighthouses and navigation buoys. He went to Sandhurst instead, an institution he 'thoroughly enjoyed'. When he joined a battalion as a junior officer, the King's Shropshire Light Infantry, he was posted to Nairobi. He arrived there to find his men were battle-hardened, having just finished fighting the Mau Mau independence movement. He flourished in the army, and volunteered to serve with the Gurkhas in Borneo, Malaysia, where they were fighting the Indonesian military. He toured the world with the army, including several spells in Northern Ireland in the early 1970s, before he eventually wound up in Colombo.

It was with some trepidation that I drove up to Holworthy's house in Bordeaux. Although I expected him to be affable, which he was, there were some rather pointed questions that I wanted to ask him, based on telegrams he had written while stationed in Sri Lanka. As it turned out, I need not have worried. Almost as soon as the interview began, he brought up KMS without prompting and reeled off details about the company that I thought he would be too guarded to divulge. After about 20 minutes I paused the interview so we could take a break, as I desperately needed to gather my thoughts in order to prepare for asking the hard questions. Again though, his nonchalance took me by surprise.

'In terms of the concerns that were being raised about the Special Task Force's activities, in terms of people being rounded up, going missing, being disappeared – do you think the Tamils were exaggerating the extent to which the Special Task Force were involved in this?'

'I don't think they were actually, no. It wasn't just the Tamils, it was the Muslims as well, and they were on the receiving end of some of this. No I think it was happening, it was in the psyche of the Sri Lankan forces at the time. They were being clobbered too, you know, left right and centre by the Tamil insurgents, and so they had a slightly different view on counter-measures to those that we in the British army have.'

'And what would some of those counter measures have involved?'

'Er, well I think they killed people basically.'

'Civilians?'

'Yes, oh yes, it was all anti-civilian. And they locked them away without any *habeas corpus*, people disappeared, there's no doubt about that. People disappeared, people were killed, that's what happens unfortunately in all these types of conflict.'

Holworthy confirmed everything he had written in the files 30 years ago. KMS's protégées were involved in systematic war crimes in Sri Lanka's Eastern Province. And worse was to come.

* * *

Amid the spiralling violence in Sri Lanka, KMS began to do much more than merely train the STF. The company reviewed the Sri Lankan military's command structure in the summer of 1985, to probe it for any weaknesses. In an attempt to shore it up, the firm provided two former SAS warrant officers, one as an operations room adviser and the other as a psychological operations adviser.[51] These would have been highly experienced special forces veterans who gave strategic advice to Sinhalese commanders as well as delving into the murky world of propaganda. The company was also increasingly involved with surveillance, and had an advisor assigned to Sri Lanka's National Intelligence Bureau, which operated in parallel to the STF. The British embassy's Alasdair MacDermott found that this KMS advisor was busily suggesting 'ways and means of improving the acquisition, collation and dissemination of information' and was 'confident that this, at least, is now slowly improving'.

As the war went on, nothing seemed off limits to the company. They were trying to persuade the army chief, Lieutenant General Weeratunga, to establish a tactical headquarters to control operations in the Northern Province and Jaffna. Reflecting on the company's growing role, MacDermott commented: 'KMS have been drawn yet further into direct control of and participation in operations', something that he warned Whitehall was 'disturbing'. He said that there was 'a real danger' if peace talks broke down and the militants launched a 'concerted push', then 'it could be they, the KMS advisers, who would provide the backbone of the resistance rather than the Sri Lankans themselves. This would inevitably become known and would cause us considerable embarrassment.'[52]

MacDermott's colleague, Richard Holworthy, had intimate access to the head of KMS in Sri Lanka, the mysterious 'Baron' – Brian Baty, alias Ken Whyte. The defence attaché was able to visit him one afternoon in August 1985, and received a thorough briefing on every aspect of the Sri Lankan security forces.[53] Baty revealed that a joint army/police jungle warfare school was being set up in the Eastern Province at Amparai, and he would draw up its syllabus and advise on training, based on his own experience running the SAS jungle wing. The pair also discussed the STF, with Holworthy receiving an overview of its camps. Locations like Batticaloa, Akkaraipattu and Komari were specifically mentioned, although there is no evidence from the records of this meeting that Holworthy made any attempt to tackle the KMS commander about what the STF was alleged to have done at these locations – the mass arrests, disappearances and fanning of intercommunal violence. This was despite Holworthy noting a 'hardening of attitudes in recent days towards Tamils'. He specifically cited the commander of the STF, Zerni Wijesuria, 'who has come in for some stick in the past for the alleged excesses of his men, was told at a recent conference that he was not being tough enough'.

When I met Holworthy at his home in Bordeaux, I showed him a copy of his minutes from that meeting. He confirmed that Baty 'was into the policy making, yes, or he was privy to it, so he was very much involved'. Later, I tried to tackle him on why he had not raised any of his concerns about the STF with Baty, but he claimed not to remember that meeting. I pushed back and asked whether he could have used his position to do more to raise concerns with the company or even stop them training the STF. 'I don't think one had any actual influence in that, they were working for their own bosses not for us, and their own bosses were in London', he rebuffed. 'They reported to them [...] They were the people to complain to.' He added: 'The high commissioner also voiced his concern about what was happening in various places, but whether that got anywhere I'm not sure.' Such was the logic of privatisation. The conduct of British military instructors in Sri Lanka was ultimately not the problem of the resident British defence attaché, because these men were working for a private company. Any issue should have been taken up with managers at a small office in Kensington, not civil servants in Whitehall or ministers in Westminster. This position was maintained despite KMS-trained personnel being involved in some of the most serious war crimes, including torture and disappearances.

By contrast, the CIA was increasingly alarmed at what Sri Lanka's police commandos were doing. American intelligence noted secretly in 1986 that the STF displayed 'greater discipline and professionalism' than the other Sri Lankan units, and its 'superior combat performance' relative to the army was attributable to the KMS training. However, this battlefield prowess was little consolation for its victims. The US embassy regarded the STF as responsible for 'most of the violence against Tamil civilians in Eastern Province. [...] a common STF tactic when fired upon while on patrol is to enter the nearest village and burn it to the ground'. Torching humans and their homes was by then undoubtedly a hallmark of the British-trained task force, something that would have horrified David Walker's fellow councillors. At one meeting, a colleague raised the issue of bonfires in Surrey, and complained that they were 'a health hazard and a great nuisance, particularly as gardens were smaller nowadays'. Walker's committee recommended that the council's leaflet on 'good neighbourliness and the lighting of bonfires be recirculated', to remind the good people of Surrey to reign in their bonfires.[54]

7

Mercenaries and Mujahideen

Six years after storming the Iranian embassy in London, Tom Morrell and Robin Horsfall were no longer in the SAS. The pair had left the British army and were now some 50 kilometres inland from Batticaloa. Surrounded by a national park, the area would one day become a safari spot for tourists eager to see wild elephants. But in 1986, Horsfall felt like he had arrived in 'the dark outback'.[1] Together with Morrell, he was stationed at an abandoned construction site in a place called Maduru Oya, 'surrounded by snakes and crocodiles', where KMS wanted the pair to run a battle school for Sri Lankan infantry.

When I first met Horsfall, some decades later, he had chosen a more upmarket location for our rendezvous. The Cabin wine bar, overlooking the platforms at Waterloo Station, is an oasis of exclusivity away from the jostling crowds of commuters. He was already sitting there when I arrived, with his SAS wings pinned smartly to the lapel of his blazer and the well-built frame that you would expect of a special forces operator. Like many SAS veterans, Horsfall had published an autobiography, mapping out the contours of his tumultuous career.[2] Although he disclosed his work in Sri Lanka in his book, he had chosen not to dwell on it there. I wanted to probe him, hoping to learn more about this episode in his career.

Horsfall grew up in an army family and spent periods living near Aldershot or on a military base in Germany. Struggling at school, he signed up to the army at age 15. For such a young recruit, he wisely asked to join the Royal Army Medical Corps or the Military Police – branches of the armed forces that would have taught him a trade useful for later in life. But both branches turned him down. Instead, he settled for his third choice, the Parachute Regiment – effectively his local unit. It was 1972, months after the Bloody Sunday massacre in Northern Ireland. Fate had chosen Horsfall to become part of the British army's most fearsome and controversial unit.

The army induction process seemed, from Horsfall's description, akin to institutionalised child abuse. When the teenagers (as young as 15) were not being screamed at, punched in the stomach or knocked unconscious by their instructors, the new recruits mercilessly bullied each other. Horsfall watched in horror as the boys in his hut subjected their weakest member to a mock hanging. They did not spare him either – one lad threw a dart into Horsfall's leg, giving him impetigo. The story of Horsfall's childhood, especially his army induction, is punctuated by senseless violence and bullying, doled out to him or others around him on the flimsiest of pretexts. The turning point was when, on the verge of quitting the army, he decided to fight back, standing up to the bullies and proving himself physically. Horsfall slowly transformed himself from being a quiet, thoughtful outsider consigned to the fringes of social circles, by acting tough enough to win the grudging respect of other soldiers. He began to excel at the army's training courses, and found he had a knack for cross-country running.

By the time Horsfall was old enough to undergo the full Parachute Regiment selection process, he was in his element – physically fit, immune to taunts and growing in self-esteem. He took the training in his stride and after weeks of living in the field and eight successful parachute jumps, was almost accepted into the elite unit. One more challenge remained – a regimental tradition called 'milling'. Each man was to fight another from his unit, to demonstrate they had the 'required level of aggression' to be a paratrooper. Horsfall was picked to fight a recruit who was six years his senior. Shaking with rage, he donned a pair of boxing gloves and pummelled his opponent to the ground, and with that earned his place in the Parachute Regiment. He had achieved his childhood dream, but would soon find that the adults he so admired were no less juvenile than the boy soldiers he had had to deal with during adolescence. While staying at a British army base in Cyprus, Horsfall was woken by being struck on the head with wooden broom handles. The attackers, two soldiers, broke his fingers, knocked him unconscious and dragged his body into a shower room. When Horsfall woke up, drenched in water and blood, his attackers were still there, grinning. They kicked and punched him some more – Horsfall could not scream for help, because they had broken his jaw – until he lost consciousness again. Eventually the assault ended and he was taken to hospital, where he spent two days passed out. It was an unprovoked attack, and in the circumstances Horsfall decided upon a fairly logical coping strategy. 'I became a psycho', he wrote. Through frequent pub brawls and other violent confrontations, he developed a reputation in his regiment as 'a nasty, aggressive brute' – a reputation that might just give others pause before ever trying to beat him senseless again.

* * *

It was a freezing winter's night when my colleague, Lou Macnamara, and I arrived in Prague for my second meeting with Horsfall. This time he would be speaking on camera, and we had come prepared with suitcases full of equipment. Prague's classic maroon trams slid us past spellbinding Gothic towers until Horsfall welcomed us into his penthouse apartment, tucked away on a quiet side street with a panoramic view of the city. A flat like this would be out of reach of most veterans in the UK, but here the pound went further, and Horsfall could enjoy his semi-retirement. The light and breezy interior design stood in stark contrast to the inside of his study, a veritable man cave, seemingly ubiquitous among veterans. One wall was lined with bookshelves, filled with tomes on military history.

Black-and-white team photos from his time in the SAS competed for the remaining space, sitting alongside other memorabilia – some of it sent by his admirers and well-wishers – including a portrait of him, and an artist's impression of the Iranian embassy siege.

Not long after that operation, Horsfall grew tired of the SAS and by age 27 found himself on what veterans call 'civvy street' – the zone inhabited by the rest of society, where people without family contacts, an established trade or professional qualifications have no clear career path to making a living. 'The skills that I learned as an infantryman weren't of value to me anywhere else but in the British army. The only civilian qualification I had was as a paramedic. And I would have chosen that career had there been jobs available for me at that time', Horsfall said. He had advised his younger brother, who also joined the army, to avoid the infantry and choose a regiment like the Royal Electrical and Mechanical Engineers, where he would at least learn a trade. Horsfall, however, could not benefit from his own hindsight. The Ministry of Defence, a behemoth of an institution and his surrogate parent during adolescence, had suddenly retreated. It was no longer paying his bills, or telling him where to live or what to do each day. The state's spear-tipped wing had abandoned him, and to fill the hole it had left behind there was only a pale imitation: the private security industry. It was the mid-1980s, Thatcherism was beginning to impose itself and in its wake privatisation appeared to be the only viable path available for SAS veterans like Horsfall. The nascent mercenary industry was about to morph from a minnow into one of the UK's boom markets, and Horsfall would have a ringside seat.

His first job was as a bodyguard for Mohamed Al-Fayed in London. Having just taken over the prestigious Harrods department store in Knightsbridge, the Egyptian millionaire was making plenty of enemies. But Horsfall told me that he soon grew bored of bodyguarding, which he found was a 'dull lifestyle'. He wanted something more exciting, and volunteered to go to work for KMS, where he would be paid more. The money was a key incentive, because he had a mortgage and one child, with another on the way. The company offered him £2,000 a month tax free, almost twice what he had earned in the SAS. 'I went where the money was', Horsfall reflected. 'I wasn't an adrenaline junkie or an adventurer, I was somebody who needed to pay the bills and these were the skills I had.' Like many veterans, he found the transition to private security was almost inevitable. 'The British army for somebody who is leaving after twelve years didn't do anything to ease me into the civilian world apart from point me in the direction of the companies that used the skills I had', he sighed. 'We're talking about 1986 here, so there weren't a huge number of jobs in conflicts that we've seen develop since then. They were few and far between. So I was quite pleased to get the work – it paid my bills.'

He was already familiar with this particular company from his time in the army. 'When I was in the Special Air Service everyone knew about KMS', he revealed. 'And when we left, many of us hoped that we would get a job with them.' Horsfall was careful to stress the close links the company had with the British establishment: 'They were nominally, nominally is not quite the right word, but they were linked to members of the British government in a deniable way.' In practice, he insisted this meant that the 'operations that they were part of were supported by the British government at the time'. Like Richard Holworthy, Robin Horsfall had taken me beyond the scope of the carefully sanitised paper trail to explain how the arrangement worked in reality.

It was straightforward for Horsfall to join the company. By 1986, their office was on Abingdon Road in South Kensington, round the corner from Harrods where Horsfall was guarding Al-Fayed. The KMS contract in Sri Lanka was well known on the private security 'circuit' and Horsfall simply sent in his CV. He was interviewed by the company's recruitment manager, Dai Prichard, a naval architect graduate, former Welsh Guardsman and the first SAS officer to go behind enemy lines during the

Falklands War. He had left the army in 1984 and was now, at the tender age of 34, trusted by KMS to vet applicants. 'I was given the job, no problems', Horsfall recalled, relieved the interview had gone smoothly. The company briefed him that he would be running a battle school in Sri Lanka, where he would be employed as a training officer and not be directly involved in the conflict.

I asked Horsfall tentatively if his work with KMS made him a mercenary, a term some in the private security industry vehemently reject. 'Well it depends on your interpretation of the word mercenary', Horsfall reasoned calmly, almost philosophically. 'If a mercenary is a person who is a paid soldier, or a voluntarily paid soldier, then every member of the British army is a mercenary. Any member of a standing army is a mercenary', he said, giving a broader definition than even Lord Diplock had done. He continued:

> If you look at a mercenary as somebody who is prepared to do anything for money, and you end up with criminal activities such as took place in the Congo, then that's another particular aspect of mercenary work. And in the time when I was working, we would refer to ourselves as 'contract' soldiers. Which is just another way of changing the word mercenary into something that means exactly the same thing. It's not the word, it's the interpretation of the word or the role that the person's been given to play. So, a standing army – full of mercenaries! A conscript army, perhaps they're not because they aren't volunteers. A mercenary involved in Angola or the Congo without any kind of government control? A bunch of bandits, which is a better description of them when they were down there. And then you've got the jobs that we had which were very strictly controlled and we would regard ourselves as contract soldiers – but mercenary would encompass all of them to some degree.

Horsfall embarked on his career as a mercenary when he arrived in Sri Lanka around February 1986, days after Tamil rebels had massacred a busload of Sinhalese passengers, in one of their worst atrocities of the year.[3] Horsfall quickly began work at the remote battle school in Maduru Oya. The training camp had opened a year earlier, after Harding and Johnson had their phone call about KMS expanding its role to the army.[4] Initially, the KMS men there retrained Sri Lanka's army commando unit, but by the time Horsfall arrived at the site the role had switched to training regular officer cadets. He was to be responsible for the final part of their training before they were deployed to operate against the Tamil militants in the north. Horsfall's immediate superior at the Sri Lankan army camp was Tom Morrell, a fellow veteran of the Iranian embassy raid. 'Initially I enjoyed the work', Horsfall remarked. 'I was training bright young men with good educations. We were teaching them counter-terrorist drills as we knew them and as we functioned in the United Kingdom with the rule of law.' He taught infantry tactics, how to clear woods, stage attacks and guard against ambushes. In other words, 'The standard things you would teach to any infantryman in any decent army throughout the world.'

There was a sign though that the company's role in Sri Lanka was about more than just training troops on firing ranges far from the front line. One day Morrell asked Horsfall to 'deploy to Jaffna for a reconnaissance'. The Tamil citadel was an all-out warzone. 'Anybody that was going to Jaffna was going to get shot at', Horsfall realised. 'Everything that was on the ground, that was on the road, was getting blown up, and there was no purpose for me to go there.' In the end, Horsfall flat-out refused to go, as did his colleagues. This minor mutiny prevented KMS from sending its instructors to this most dangerous corner of the country, but the company would have its way in the end.

* * *

Jihadi for Hire

May 1986

While Horsfall was carving out a new career for himself in Sri Lanka, another British army veteran was desperately trying to make a living. Captain Anthony Michael Knowles started life with a bright future, studying at the renowned Royal Military Academy in Sandhurst before joining the prestigious Household Cavalry.[5] He aspired to be in the elite and trained with Britain's special forces, until a car accident in 1983 forced him to drop out of SAS selections. It was a severe blow for an extremely fit man in his early twenties. Despite his injuries, he still wanted to play a decisive role in international affairs. The Soviet invasion and occupation of Afghanistan was something that particularly concerned him, in common with many conservatives of that era, who saw it as a chance to give the USSR its own Vietnam. By 1986, Knowles had 'for some time been trying to do something – anything – about Afghanistan', a recently declassified telegram at the UK National Archives reveals. Captain Knowles was eager to persuade a US diplomat at their London embassy to help him, 'and she has proved more sympathetic than most'. Others were more wary of this injured veteran, including dissidents in London. One group of exiles, the Afghanistan Support Committee, believed that Knowles was 'unreliable and more likely to prove a liability to any organisation who took him on to work for the Afghanistan cause. The [British] Embassy in Islamabad share that view.'

Although the odds seemed stacked against this plucky veteran, the same telegram states that Knowles had 'come up with a scheme under which KMS (the same Keenie Meenie Services who help the Sri Lankan Government) would train the Afghan resistance in demolition techniques etc, inside Afghanistan'. While this might sound remarkable by post-9/11 standards, it was something that many right-wingers wanted to pursue back then. Knowles had relayed his scheme to the US embassy, where Robin Raphel, an ex-CIA staffer, had 'passed the proposal to "the other side" in Washington. She had no idea whether the proposal would be taken up.' This 'other side' could be a reference to the US State Department, the CIA or the Pentagon. Whichever it was, at that time KMS would have been well known to America's national security elite, given David Walker's work with Oliver North in Nicaragua. Britain's Foreign Office appeared more concerned about the proposal, commenting:

> It would be unfortunate if KMS were given a contract of this nature. Already in Sri Lanka, they are regarded as agents of HMG and it would soon become fairly generally known that they were acting inside Afghanistan, which could become an embarrassment to us. And if, of course, any of the instructors were taken prisoner by the regime/Soviet forces, the publicity would be very unfavourable.

The civil servant concluded: 'If you agree, I think it would be worth [one line redacted] to try to discourage anybody from hiring KMS for the training Mr Knowles is suggesting.'[6] The telegram was then forwarded to another part of the British government, presumably the intelligence services, because the recipient's name is redacted. The sender signed off by asking the recipient to 'pass it on to the appropriate quarters to consider whether any further action should and can be taken'.[7] The file then ends, and KMS's Afghan paper trail runs cold.

* * *

It has long been rumoured that British mercenaries, and specifically KMS, assisted the Afghan Mujahideen against the Soviet occupation in the 1980s. This terse telegram above is, however, the only documentary evidence of this relationship that I have found, and even then it is inconclusive at best. However, the telegram's tenor fits into a wider pattern of MI6 and CIA involvement with the Mujahideen, and with the *modus operandi* adopted by KMS in Nicaragua. As early as June 1979, when Soviet tanks rolled into Afghanistan at the request of the country's communist government under Nur Mohammad Taraki, the UK Foreign Office was paying close attention to any Britons who managed to liaise with Afghan opposition groups. The occupation led to a sudden surge in colourful characters asking the anti-communist block for help, which caused a moment of introspection in Whitehall about what to call such allies. The opposition to the Soviet occupation ranged from Afghan Maoists to Islamists, and the Foreign Office's South Asia Department (which readily referred to Tamil fighters as terrorists and extremists) lamented: 'Using the term "rebels" to describe the Afghans fighting against Russian forces carries the connotation that they are people fighting against the duly constituted authority of the country.' They noted agreeably how other branches of the Foreign Office spoke of nationalists and resistance, and the diplomats preferred terminology such as 'guerrillas' or 'patriots' instead of rebels, which they were dismayed to find was used by the Foreign Office-funded BBC World Service.[8]

Lord Carrington, then Britain's foreign secretary, went further and said that the term 'freedom fighters' would seem more appropriate.[9] In a similar vein, the UK's permanent representative to NATO told the Cabinet Office that 'it would be psychologically better for us, even in Secret documents, to speak of the "resistance movement", "national liberation movement", "freedom fighters" or some such upbeat phrase'.[10] As the war went on, Thatcher's foreign policy chief then observed that Afghan insurgents had plenty of small arms but lacked the heavier weapons necessary for large-scale operations against Soviet tanks and helicopter gunships. The foreign secretary suggested ominously:

I have in mind the sort of man-portable missiles that infantry use against low-flying aircraft, infantry anti-tank weapons, and anti-tank mines. It seems to me that this, though a tricky area, is one where the Afghan liberation movement deserves a measure of support, and one where the injection of some slight assistance may pay substantial dividends. Perhaps our officials could take this further.[11]

The British government's files from this period had been meticulously censored before they reached the National Archives, expunging any reference to the intelligence agencies or special forces, making it difficult for the public to know how far British officials took Lord Carrington's suggestion. However, former members of the SAS have been more forthcoming about the UK's work with the Mujahideen. Ken Connor, an ex-SAS warrant officer who spent years in the elite unit, claims that in 1981 MI6 opted to work with a Mujahideen faction led by the so-called 'Lion of Panjshir', Ahmed Shah Mahsood, who was later assassinated by Bin Laden two days prior to 9/11. Connor's claim is supported by another author, who wrote: 'MI6, which operated out of a small windowless office in Britain's Islamabad embassy, made contact with Mahsood early in the war and provided him with money, a few weapons, and some communications equipment.' MI6 also taught English to some of his key aides, including his foreign policy guru.[12]

By 1982, this programme of covert British assistance saw an MI6 team slip into Afghanistan from Pakistan, using pack-horses to carry supplies and equipment. However, they were ambushed on the way out, the horses were killed and their rucksacks captured. British passports were paraded to a press conference, and Connor says this exposure led MI6 to rely increasingly on recently retired SAS members to do this work instead of serving British spies. This would suggest that companies like Keenie Meenie, or even KMS itself, was working with the Mujahideen as early as 1982. This informal arrangement would have been familiar to Colonel Johnson from his pre-KMS days running a mercenary war in Yemen, and chimes with Horsfall's claim that the company was 'very well connected' with both MI6 and the SAS.

Although Connor does not specify which company they worked for, he claims that at some point in the early 1980s ex-SAS warrant officers trained Mujahideen inside Afghanistan in communications and command techniques. One winter, Connor says, junior officers from Mahsood's faction were taken out of Afghanistan by two ex-SAS non-commissioned officers and trained in the Middle East, as well as at a barn in England's Home Counties, and also in the north of England, in the Scottish borders and on Scotland's west coast. Sessions included live ammunition, aggressive tactics, planning operations, the use of explosives and fire control of heavy weaponry such as mortars and artillery. Connor claims that these Mujahideen were taken on helicopter flights to see how hard it is to spot human targets from a couple of hundred feet. They also practised entry into airbases and how to lay anti-aircraft mines on the centre of runways. As if this was not enough, they were taught how to lure ground-attack aircraft into narrow valleys, where they could be destroyed by crossfire from valley walls, and mount 'linear anti-armour ambushes'. Equipment for this scheme was supplied by a third-party sponsor, according to Connor. All these techniques would have been familiar to David Walker and his men in KMS, and sound similar to what he wanted Oliver North's Contras to attempt.

Connor does give more precise details about one operation, which allegedly took place in autumn 1982. He says MI6 arranged for two ex-SAS men in London to meet a CIA operative who wanted their advice on how Afghan guerrillas could destroy Soviet MIG aircraft on the ground in Afghanistan. The retired SAS soldiers drew up a plan, and a month later Connor claims that 23 out of the 24 MIGs were destroyed. Another source suggests that *serving* SAS soldiers were taking an even more active role with the Mujahideen shortly after this incident. In 1983, the CIA's new Afghan Task Force chief, Gustav Avrakotos, visited MI6 headquarters in London to introduce himself. He quickly established two things. MI6 was 'virtually bankrupt', but operated with far fewer legal limits than the CIA, whose dirty war in Nicaragua was attracting the ire of the US Congress. The CIA man was also impressed at Britain's knowledge of Afghanistan, a region they had been operating in since the nineteenth century. This insight had allowed them to recognise the importance of Mahsood, who controlled the strategically vital Panjshir Valley, which was a bottleneck for Soviet supply lines to Kabul.

The CIA's Afghan chief wanted to meet MI6's Mahsood expert, who he claims turned out to be a 'young, blond SAS guerrilla-warfare expert with the peculiar nickname Awk', owing to a grunting noise he made during battle. This SAS man had just returned from three months in Afghanistan with two other SAS colleagues, and it had been an eventful tour. They had travelled with Mahsood's men, and awoke one night to hear one of Mahsood's lieutenants anally raping a Soviet prisoner. This lieutenant would later save the SAS trio when they were caught in a Soviet ambush, by charging out into the open and drawing helicopter fire away from them, allowing the British commandos to survive. The CIA man was impressed by this gruesome tale, and wanted to send the MI6/SAS team straight back into the field. The only thing holding the Brits back was a lack of mine detectors, which meant Mahsood's men kept getting maimed. The CIA agreed to supply 25 mine detectors immediately, and were horrified that MI6 was too broke to even afford this gear (which only cost around $7,500).[13]

It is conceivable that the visiting CIA man was so wide-eyed that he missed nuances about who the SAS men really were, and whether they were serving members of the British army or guns for hire. For the CIA, using MI6 assets had significant benefits. Avrakotos described Thatcher as 'to the right of Attila the Hun', and marvelled at the lack of legal constraints on MI6. 'They had a willingness to do jobs I couldn't touch', the CIA chief exclaimed. 'They basically took care of the "How to Kill People Department"'.[14]

By the mid-1980s, the war in Afghanistan was attracting an increasing number of mavericks: English arms dealers, hawkish Conservative politicians and former SAS soldiers were all keen to get a piece of the action against communism, regardless of whose hands their guns and money ended up with. It was a simple case of my enemy's enemy is my friend, a maxim with consequences for the region and world that we are still reaping today. The level of British involvement remains highly classified, and if it is ever made public will likely be staggering. Although the Foreign Office appeared apprehensive at becoming too heavily involved in Afghanistan, its sister organisation, MI6, seemed to have no such reservations. In addition to its work outlined above, MI6 is said to have provided 'magnetic depth charges to attack bridge pylons', including one that connected Afghanistan with Soviet-controlled Uzbekistan.[15]

It is in this context then that Captain Knowles' tentative approach to the US embassy in London in May 1986 should be understood. Knowles was clearly a maverick, but as such he would have fitted right in with the wider attempt to support the Mujahideen. If he was really working for KMS, then the company was clearly eager to win a lucrative new contract, and repeat Walker's efforts with the Contras. I made freedom of information requests to the US State Department, Pentagon and CIA about Knowles, to try to find out what happened after his call to the US embassy in London. Unfortunately, these US agencies are yet to declassify any of their documents about him.

Whatever happened to Knowles' Afghan initiative, he has recently been in trouble with the law. In 2016, he was jailed for four years after admitting to fraud at Liverpool Crown Court. The old captain had taken £140,000 from the tax man, and tried to falsely claim another £1.8 million in VAT repayments. He reportedly spent the money on 'expensive jewellery and stays in top London hotel Claridges', claiming that the ruler of Dubai had lent his business £25 million when a revenue officer became suspicious. He had also clocked up a conviction for theft in 2008, after 'falsely claiming £69,000 on his War Pension by telling the Ministry of Defence he was incapable of work, despite running three companies'. In 2016, he attended his VAT fraud trial in 'a wheelchair and breathing with the use of an oxygen machine', claiming that he needed '24 hour care from his wife and daughter and had limited memory of the offences'. However, the judge was not impressed. 'You have presented as a severely disabled person', he noted sternly. 'I have read medical reports and I can find little support for that presentation.' Knowles was deemed fit enough to have committed serious fraud and enjoyed spending the substantial proceeds.

Be that as it may, the sentencing clearly distressed his wife and daughter, who a reporter said 'sobbed as he was led to the cells'. Upon learning of his incarceration, I had hoped to speak to Knowles about his work with KMS in Afghanistan. However, up to a dozen messages to his lawyer's office in 2018 were not returned, and the Prison Service said its 'records hold no trace of this person being held in prison custody at present'.[16] Again, the trail ran cold.

* * *

Carlisle, 1986

With Captain Knowles allegedly soliciting work for KMS in Afghanistan, the company simultaneously set about recruiting another veteran to help expand its operations in Sri Lanka. Brought up in the back streets of post-war Bristol, and beaten by an abusive father, Tim Smith joined the British army as soon as he was old enough to leave home, at age 15. He went on to serve in the Royal Tank Regiment, and rose quite quickly through the ranks to become a sergeant by his mid-twenties. Seeking new challenges, he learned to fly helicopters and became a pilot in the army air corps, serving as far afield as Hong Kong. Closer to home, he flew missions in Northern Ireland during the Troubles, dodging bullets from the IRA as he patrolled the border, or giving aerial cover to troops as they cleared up the carnage of culvert bombings. He was awarded a Queen's Commendation for 'valuable service in the air' in 1982.[17] After 22 years in the army, and two failed marriages, Smith called it a day. Like many veterans of his generation, who left the army as Thatcherism was taking hold, he went into the private security industry. Smith spent the next two years training Saddam Hussein's young pilots to fly helicopters, for an outfit called AirTran. (Iraq was at that time Britain's ally in the war with revolutionary Iran.) He tried to settle down with another woman, Eileen, who fell pregnant, but Smith grew restless. He was preoccupied with earning enough money to repair their home in Carlisle, and provide for his 13-year-old son. It reached the point where the roof was leaking, and an old friend from the world of private security was suddenly calling.

Smith retold this turning point in his memoirs, which were based on his diary. It provides a raw insight into his train of thought. When he answered the phone that day, the line was so bad that it sounded like he had 'his head in his dustbin'. An old friend was claiming to be earning 'easy' money, tax free, working as a flying instructor in Sri Lanka. They taught when it was coolest, in the early morning and late afternoon, punctu-ated by a siesta at midday. The nights were spent sipping Arak on the veranda or at the capital's best hotels. Smith was sold already, and only had one question – what did KMS stand for? 'Keenie Meenie Services, I think its Maori, or something like that', his friend blustered. Smith was so desperate for a change that he 'wouldn't have given a toss if they were green-skinned, double-headed, baby-eating monsters', so long as the pay was reasonable. It turned out Smith would earn the same amount as the North Sea pilots who ferried crews out to gas rigs, only in Sri Lanka the weather would be a lot warmer.

Catching the sleeper train from his home in Carlisle, this wiry old pilot with a trimmed moustache and short grey hair soon arrived at Abingdon Road. The KMS headquarters had a scruffy, unassuming facade, guarded discreetly by an intercom. Once buzzed inside, he was interrogated by the same veteran that had grilled Robin Horsfall – Dai Prichard. He probed Smith's background extensively, before taking him out to lunch at an exclusive London club, crammed with old officers, city gentlemen and rugby players. Smith sensed they were Parachute Regiment veterans, with the 'stiff upper lip of the Old Boy Network'.[18] Before long, Smith bumped into someone he knew from his army days, kitted out in a tweed suit. The pair made polite small talk, before his acquaintance flitted back into the crowd. Smith was being vetted surreptitiously. When he sat down for lunch with the KMS recruitment staff, a man in a grey suit, apparently from the Ministry of Defence dined with them and spoke cryptically about a potential contract for the company abroad. Smith concluded from this unconventional job interview that KMS had 'some sort of Government blessing, and a tenuous connection with the Ministry of Defence'.

By the afternoon, Smith was back in Abingdon Road being given what felt like a full-scale military briefing. As the overhead projector whirred, slides of STF recruits at Katukurunda flashed on the screen. Cuttings from the *Daily Telegraph* provided some background to the situation in Sri Lanka, and any references to mercenaries were 'poo-pooed with a muttered, "We're a security and training organisation"'.[19] Even if Smith were following the papers avidly, which he was not, none of this would have bothered him. By now, he had pinned his hopes on working for KMS. He passed the company's rigorous vetting, which involved them quizzing locals at pubs in Carlisle without his knowledge, where they managed to establish that his wife was pregnant. All that remained was for him to sign the contract, which was written according to Cayman Island law, the opaque tax haven the company had chosen to funnel its affairs through. Smith also had to open a bank account with Morgan Grenfell in the Channel Islands, another tax haven.[20] After all this, he was finally ready to go, just a week before Eileen was due to give birth. Regardless of the due date, he pushed on. Smith spent his last night in England with an ex-girlfriend staying at the Heathrow Penta, a brutalist airport hotel.[21] He had no idea what else was in store.

*   *   *

'Some of the officer cadets came back and said that in spite of all the training that we'd given them they were being asked to necklace prisoners', KMS instructor Robin Horsfall told me sombrely. 'That means you put a burning rubber tyre over somebody and burn them to death, or use it as a form of torture.' It was a tactic countless Tamils had told me about as well. 'It's terrible,' Horsfall continued, 'burning somebody alive essentially to punish them, to hurt them, to create shock, that's what they were being asked to be involved with as soldiers and they weren't happy about it. They were carrying out killings that they were being ordered to carry out by their senior officers.' Horsfall did not witness these atrocities directly, but he heard about them from the young officers, 'who came back to me and were very distressed about it'.

It is difficult to tell precisely which massacres Horsfall was referring to, because by this point in the war such incidents were frequent. He recalls arriving in Sri Lanka around February 1986. To give a sense of some massacres that Horsfall's trainees may have participated in, it is worth recalling an incident on 19 March 1986. Sri Lankan troops drove into Eeddimurn-chan, a farming village in Vavuniya district, around half-past four in the afternoon. They were accompanied by Sinhalese settlers – convicts who were being used to colonise Tamil areas. Homes were looted and raised to the ground, and seven local civilians were shot dead. Early the next day, at another village (Nedunkerni), the same troops started 'shooting everyone including old people and children. 20 people were killed in the two days of violence. Property worth hundreds of thousands of rupees was also damaged', a human rights report states.[22] Throughout the incident, a Sri Lankan Air Force (SLAF) helicopter provided cover for the operation, strafing surrounding houses and damaging buildings up to three kilometres away. The survivors were so frightened that they hid in the surrounding jungle, where they secretly buried their dead relatives.

'What really was the tipping point was when I got back to Colombo for a weekend', Horsfall presaged. A South African KMS pilot began sharing his war stories. A Sri Lankan vehicle had been blown up by a landmine and the pilot scrambled to provide air cover. 'As he flew over the area his door gunner opened up and shot every man, woman, child and animal that he could get his eyes on', Horsfall recalled. 'He flew his helicopter much higher to prevent him from doing that.' It was a grim window into how closely Horsfall's colleagues were brushing with war crimes, and the lack of consequences. 'There were no recriminations', he remarked. 'It was considered to be perfectly acceptable by those people that were in charge of that group.'

This anecdote helped Horsfall make up his mind. 'For me it was a tipping point. There was nothing I could do. I was a fairly young man, I was coming up to 30, I had no authority. There were no press in the north of the country, there was nobody you could take the story to. So I just simply decided to go home.' After barely four months in Sri Lanka, Horsfall had had enough and compared the island's ethnic conflict to the Holocaust. He had certainly experienced a stark introduction to life in the private security industry. 'I'd only been out of the army about 15 months so it was a world that was very unusual to me and this was my first real civilian-military job', Horsfall told me. To exit the country, he pretended to be sick and refused to return to Maduru Oya. 'I created the circumstances to be called into the office, gave the wrong answers and was put on a plane home', he said frankly, underplaying the significance of his rebellion.

At the office, he was met by Brian Baty, the company's chief in Colombo and his former superior in the SAS. Horsfall refused to be deferential, saying he was 'obtuse, or stroppy – not kowtowing to his authority, not giving it yes sir, no sir, whatever you want sir, I'm really sorry sir'.

Horsfall left Sri Lanka on 3 May 1986, just as a bomb destroyed an Air Lanka civilian plane on the runway at Colombo airport. It was a dramatic ending to a harrowing stage of his life. 'Inside I was pretty upset about the whole affair', he said. For KMS though, the situation was more straightforward. The company had a vacancy to fill. Less than a fortnight later, around 15 May, and with Captain Knowles trying to expand the company's work to Afghanistan, a seriously hungover Tim Smith touched down in Colombo. The humid airport was crawling with armed police, fearing another Tamil Tiger attack. As he collected his baggage, a tanned athletic figure with wavy greying hair sauntered up to him and began to help. It was Brian Baty, the KMS team leader. As a sign of his seniority, the pair were then able to exit the airport, bypassing customs, immigration and security.[23] A car took them the short journey to the military side of Katunayaka airport. It was home to the SLAF's Number 4 Squadron. Surrounded by coconut trees, the base was fitted out with familiar British-style aircraft hangars, relics from the colonial era. Baty promptly introduced him to Squadron Leader Oliver Ranasinghe, a Sinhalese air force pilot who was 'tall and slim, almost elegant, the very picture of a well tanned British cavalry officer'.[24] Smith was given the honorary rank of captain, a significant promotion from his career with the UK military. However, his expectations of the high-life were immediately dashed upon being shown his accommodation – 'a ramshackle heap of wood and wriggly tin ... the place was a wreck, a complete and utter bloody shambles'.[25] Far worse was in store.

8

The English Pilot

The charge sheet against KMS contains a mixture of moral, ethical and legal violations. In Oman, the company protected a tyrant, much to the chagrin of the sizeable number of Dhufaris and Omanis who wanted control of their own political destiny. In Afghanistan, KMS allegedly passed on demolition techniques to the Mujahideen, and it is unclear if any of these methods later fell into the wrong hands. In Nicaragua, David Walker is directly accused of sabotaging both military and civilian installations, including a hospital, a crime under Nicaraguan law. In Sri Lanka, the company was clearly on the side of the oppressor. For many Tamils, this alone will be enough to condemn them. KMS would argue that its involvement was solely limited to training and accordingly it could not be culpable for the crimes its trainees went on to commit. This justification may have been acceptable, had the company not continued its training after countless atrocities. Their defence is further weakened by the fact that KMS extended its coaching role to other branches of the security forces that were known to relish massacres, and provided strategic advice at the highest level of the country's security forces.

But the company's work in Sri Lanka did not end with their police, army and intelligence agencies. To stop at that would have left the Colombo government dangerously exposed to the Tamil independence movement, whose young men and women were determined to liberate their homeland from Sinhalese subjugation. These rebels were well armed, well organised and had nothing to lose. They moved seamlessly among their people, or as Mao would say, as easily 'as a fish swims in the sea'. And even far from the coast, this nation of seafarers and fishermen, who are said to have invented the catamaran, proved that they could steam around in lightweight fibreglass skiffs, with several high-power outboard engines bolted to the stern. These craft could ferry supplies in from around the Indian Ocean or skirmish with the Sri Lankan navy.

In fact, the Tamil rebels were so powerful that by the summer of 1985 the Sri Lankan government was engaging in peace talks. The negotiations took place at a summit in Thimpu, the capital of the Himalayan nation of Bhutan. Rival Tamil factions, from the LTTE to PLOTE, EROS to the EPRLF, TELO[1] to TULF, had been corralled by community organisers into putting on an unprecedented show of unity. One civil society activist, Varai-muttu Varadakumar, worked tirelessly on this task, missing the birth of his child. And yet he succeeded in drafting a historic set of demands for Tamil self-determination that the ad hoc Eelam National Liberation Front (ENLF) could rally around, and future generations would regard as a watershed moment. Poignantly, the ENLF's conditions for a ceasefire agreement also included a demand that the Sri Lankan government 'send back home "British SAS trained mercenaries"'.[2] This summit truly could have been a tipping point in the conflict, which brought an end to the fighting and gave the Tamils a state of their own. Militarily, they held a slight advantage over the Sri Lankan government forces.

However, the Sri Lankan government had almost no intention of complying with any of their demands, and used the lull in fighting to enlist additional KMS assistance. To contain and crush the burgeoning Tamil revolution, Colombo needed control of the skies – over both the jungles and the sea. The air was one terrain where the Tigers could not compete – the cost and complexity of even the most basic military aircraft was beyond their reach. Helicopters had been used decisively against the young JVP rebels in 1971, when Bell Huey helicopters were hastily delivered to Colombo. Over the next decade, this fleet gradually grew in size, although there was a shortage of skilled pilots who were able to operate them. The Huey was a workhorse, synonymous with the Vietnam war. It could fly soldiers into battle, evacuate the wounded, and resupply isolated bases with food. A machine gun was strapped to the side door, operated by a crewman, and some versions had podguns for the pilot to fire.

KMS was about to demonstrate that it was as adept at captaining helicopter gunships over Sri Lanka's jungles as it was at sabotaging them on a helipad or downing them with surface-to-air missiles. It would be this aspect of KMS's operations that embroiled its staff most directly in war crimes, as well as the moral and political transgression of wrecking a peace process and ultimately denying a people their homeland. The escalation of KMS's operations in Sri Lanka also had profound implications for the British government. When the prospect of KMS pilots was first floated in June 1985, Whitehall appeared somewhat alarmed – or at least wanted to register its concerns periodically for the record, as it had already done around the start of the year when Harding spoke to Colonel Johnson. This time, diplomats got wind of the new scheme from a visiting salesman in Colombo, a retired army helicopter pilot who mentioned that KMS was attempting to recruit his old colleagues in the UK. The high commissioner warned London: 'we are getting dangerously close to KMS acting in an operational rather than purely training role'.[3] Curiously though, it took the Foreign Office a fortnight to respond. 'This is disturbing', the department remarked. The envoy was instructed to confirm the development with Brian Baty, and then Whitehall's main interlocutor on this matter, Harding, had a word with Colonel Johnson about whether the company really was employing pilots.[4]

And so it was amid the delicate summer ceasefire that David Walker and Brian Baty waltzed in to the British High Commission in Colombo and confirmed that they were providing eight pilots: five helicopter instructors and three fixed-wing aircraft teachers.

These foreign airmen represented a significant addition to the SLAF, which had 20 helicopters but only six of its own pilots willing to fly over combat zones. Walker refused to guarantee to the British envoy that his KMS pilots would avoid operational areas, although he said they 'always fly with SLAF co-pilots, and would not man weapon systems'.[5] And with that, the conversation quickly turned to other matters. Baty was scathing about the ceasefire, saying it had 'produced inertia throughout the army',[6] and that the only actor pushing for a military solution was Athulathmudali, the national security minister whose office Baty most probably worked next to. The minister had ordered all training and strategic planning to continue, with KMS advising the security forces to plan counter-terrorist operations in detail in case the ceasefire broke down. Baty seemed actively hostile to the ceasefire, and felt the military was 'putting too much faith' in the peace talks. The high commissioner did try to quiz Baty about alleged STF abuses, including a case where 40 Tamils had been forced to dig their own graves before the commandos executed them. Baty defended his protégés, insisting that 'the majority of the reports were not true' and that he trusted the integrity of the unit's commandant.

Within the Foreign Office, the company continued to divide opinions, hampering how the department responded to the prospect of KMS pilots. One diplomat argued that their effectiveness was 'debatable', citing its dubious track record of training the STF. Furthermore, recruiting KMS pilots might suggest President Jayewardene was preparing to resume military operations, which called into question his desire to reach a negotiated settlement at the peace talks in Bhutan. The foreign secretary, Tory MP Geoffrey Howe, was particularly pessimistic: 'we could now be on a slippery slope', he remarked.[7] The company's harshest critics within Whitehall wanted to 'make life very uncomfortable for them', although they struggled to identify specific sanctions that could be implemented.[8] The carefully constructed vacuum that flowed from Diplock's 1976 review of mercenary legislation continued to leave quizzical officials powerless to prevent British veterans doing as they pleased in Sri Lanka.

Amid the delicate ceasefire, one of Britain's highest-ranking diplomats met the KMS leadership in London. Harding confronted Colonel Johnson and Major Walker, who confirmed they had recruited four British pilots, one American and three Rhodesians. The pilots would be paid and controlled by KMS Ltd, 'through a front company in the West Indies', presumably the Cayman Islands. It transpired that Sri Lanka's defence secretary, General Attygalle, had offered the job to KMS after a 'cowboy' Australian operator had proven unsatisfactory. Indeed, Colonel Johnson argued that it was preferable KMS did this work rather than 'a mixed bag of mercenaries'.[9] Harding duly conveyed the foreign secretary's concerns that these pilots would fly in operational areas, their potential impact on the ceasefire and the likely backlash in New Delhi[10] However, the colonel was unphased and suggested Whitehall should take the matter up directly with the Sri Lankan authorities.

The colonel's 'obduracy' riled the foreign secretary, who ordered Britain's ambassador to meet with Sri Lanka's hawkish national security minister, Lalith Athulathmudali, and raise the issue of KMS pilots directly with him. At first, Athulathmudali tried to claim the pilots would only be required for a very short period of time, but eventually he admitted they had signed one-year contracts which included the possibility of operational flying. The minister tried to placate the ambassador, John Stewart, by saying KMS would probably have trained a sufficient number of local pilots before the ceasefire broke down. However, Stewart was already sceptical of KMS and regarded their latest assurances as 'nonsense', predicting the ceasefire would only last another eight weeks which was not enough time to train pilots from scratch.[11] The air force was so short-staffed that Stewart expected the minister would 'rashly commit' KMS pilots to combat operations if the ceasefire collapsed.

Whitehall was not satisfied by Athulathmudali's response and now instructed the envoy to raise the matter with President Jayewardene himself.[12] The sincerity of Whitehall's concerns was undermined, however, by the fact that, while expressing displeasure, it was simultaneously awarding the Sri Lankan government more development aid to build a power station in Samanalawewa – a Sinhalese area in the south of the country.[13] There was also a fundamental flaw in Whitehall's approach. The ambassador was instructed to begin his meeting with Sri Lanka's president by praising Jayewardene for taking part in the peace talks and reassuring him that it was 'self evident that the Sri Lankan government cannot abandon the principle of a unitary state'. In other words, he did not need to succumb to Tamil demands for Eelam. Despite all the bloodshed in Sri Lanka, the British government refused to accept that the single state structure it had imposed on the island at independence was at the root of the conflict. Whitehall was giving President Jayewardene a green light to deny the Tamil demand for self-determination, even as the rebels held the military balance of power. This approach would inevitably scupper the peace talks. Why should the Tamil revolutionaries give up on a separate state when it was within their grasp?

Britain's envoy thus embarked on his doomed task. He saw President Jayewardene one evening, and the conversation eventually turned to KMS. At first, Jayewardene insisted that KMS employees would only be used for training. Stewart then revealed that a senior KMS employee had told him the pilots could fly on operations. The president had been caught out, and now promised he would immediately instruct that the company's pilots could not take part in combat in any circumstances. However, Stewart was unimpressed, and increasingly cynical of Jayewardene's intentions. 'The Sri Lankan side has, in agreeing to negotiations with the Tamil separatists, been largely motivated by their need for a breathing period', Stewart told London. 'They believe that during the ceasefire they will be able to train and equip their armed forces to a stage where they will be capable of dealing with what will almost certainly become a civil war.' In this context, the KMS pilots would clearly play a decisive role. Stewart could see the writing on the wall. 'Despite the president's unequivocal promise I am not convinced that, at need, the KMS will not be used in an operational role', he continued to warn. Despite this gloomy forecast, the cloud had a silver lining: 'The value of his promise lies in the fact that we can now truthfully maintain that the Sri Lankans promised that the pilots would not go outside their training role nor be employed on operations', Stewart concluded. This card would be kept up Whitehall's sleeve, ready for ministers to deal in Parliament one day should nosy MPs ask questions about white men with English accents flying gunships in Sri Lanka.

* * *

Regardless of Whitehall's apparent reticence towards KMS pilots, the Foreign Office team in Colombo continued to have close links with the company. By August 1985, a month after Sir William Harding had warned KMS about providing pilots to Sri Lanka, Britain's defence attaché, Lieutenant Colonel Richard Holworthy, sat down with the company's commander in Colombo, Brian Baty, to receive a briefing on the civil war. Baty said five KMS helicopter pilots had arrived in Sri Lanka, and two fixed-wing pilots. Their eighth instructor would be posted to China Bay, a major air base in Trincomalee, in late September. Baty went on to describe the pilots' daily routine: a seven- or eight-hour day ferrying rations around in the helicopters, and bringing in fresh soldiers. The resupply operations were so demanding that there was very little time for teaching. 'They are not doing the task for which they were recruited, which is training the Sri Lankan Air Force in tactical flying', Holworthy learned from Baty. The KMS pilots were installed in influential positions, with one man flying the coordinating officer for Trincomalee, Commodore Sujith Jayasuria, around.[14]

The British envoy also monitored the whereabouts of KMS pilots, spotting two on a visit to the Palaly airbase in Jaffna in mid-August. They were flying the Bell helicopters and co-piloting Avro transport aircraft.[15] This military build-up had not escaped the attention of the Tamil militants, who were 'all aware of the presence of these pilots' by August, when the ceasefire negotiations were still going on.[16] The alarm bells sounded briefly in mid-September, when the British High Commission reported that a helicopter was forced to land with engine failure during operations to retake Trincomalee. 'We do not know yet if the pilot was a KMS employee', an anxious diplomat relayed.[17] By now the ceasefire had broken down, and the war was once again raging.

Amid the constant firefights, one poignant incident seems to be missing from the British archives. A fortnight after the helicopter crash-landed, other choppers made a very controlled and deliberate descent into Piramanthanaru, a Tamil farming village. There, in the early hours of a misty morning on 2 October 1985, Sri Lankan soldiers streamed out of helicopters and into irrigation channels. Tamil witnesses insist that while this was going on, a white-faced pilot waited beside one of the helicopters. He watched as the troops he had delivered conducted the round-up and execution of 16 male civilians, aged between 17 and 33. Blood poured from a man's ears, others hung upside down from a tree as they were waterboarded. Homes were torched, and a corpse dropped from a helicopter and left to fester in the undergrowth until the dead man's sister smelt the stench.

Given that the air force was heavily reliant on KMS pilots at this point in the conflict, it is almost certain that the white man in Piramanthanaru that fateful morning was a company employee. The intrepid Scottish diplomat, Alasdair MacDermott, visited the north-east a month after the massacre and observed that KMS 'fly helicopters regularly around the island including to Trincomalee (a "white" pilot flew a group of local journalists there last week)'.[18] In another telegram, he told Whitehall's

Cabinet Office: 'KMS appear to be becoming more and more closely involved in the conflict and we believe that it is only a matter of time before they assume some form of combat role however limited it might be.' MacDermott continued to sketch out the extent of KMS involvement in the conflict, reporting that the Australian ambassador had recently been flown around Sri Lanka in a helicopter gunship piloted by an American, who was almost certainly one of the eight KMS airmen. MacDermott warned that KMS members were frequently seen in hotel bars in Colombo and their identities were well known among the British expats.[19] He seemed most concerned about Westerners seeing the pilots, and slightly less fussed about whether these same airmen were flying local troops into battle, so long as it stayed out of sight.

Despite the company's increasingly high profile in Sri Lanka, the people of Elmbridge had little idea that one of their elected councillors was embroiled in the massacre of Tamil villagers. In 1985, as KMS began flying helicopter gunships in Sri Lanka, Councillor David Walker joined Elmbridge Council's Aircraft Noise Advisory Committee. It was opposed to a fifth terminal at Heathrow Airport, any expansion at Stansted Airport or indeed any more flights per year at all. They were also concerned about helicopter noise in London.

*　*　*

Whitehall, November 1985

As the year drew to a close, a group of civil servants finally began to conduct some due diligence on KMS and its connections to the British government, fearful that the company's role in Sri Lanka would escalate further. Until now, although KMS had attracted some disquiet in Whitehall's corridors, these concerns had quickly been rebuffed by supporters such as Harding, who effectively saw off any attempts to investigate the company.

As a sign of this continued lack of consensus, the Ministry of Defence produced two reports that reached radically different conclusions. The department's arms-dealing wing, the Defence Sales Organisation (DSO), confidently claimed the company was estranged from the British state. However, the Ministry of Defence's more secretive Defence Intelligence division felt that KMS could effectively be considered an 'unofficial arm' of the UK government.

For its part, the DSO held a dossier on all of Britain's private security firms, which included a sketchy one-pager on KMS. It listed the company's address as Morgan Grenfell Trust, a bank that had been registered at 12 Dumaresq Street in St Helier, Jersey, since 1980. The DSO said the company's origins were unknown, but noted correctly that there was a 'possible connection with Saladin Services' and another firm called Lawn West. They speculated that KMS may be an 'undercover offshoot' of an insurance firm called Thomas Nelson. Under a heading 'line of business', an official wrote ominously: 'reported to be special operations'. The officials also believed KMS had supplied bodyguards during the 1984 Olympics in Los Angeles.

More importantly, the firm had several points of contacts with the DSO, including in its marketing services division and the British Army Equipment Exhibition, a biannual arms fair. Regardless of these connections to Britain's defence ministry, the report's author concluded that there was 'very little known [about KMS], but name crops up sometimes'. The DSO's regional marketing directorate believed the company 'should be black-listed for e.g. activities in Sri Lanka', and KMS was given a category 'C' rating – the lowest possible, suggesting that there was by now considerable dissent within Whitehall about how to handle this company. An accompanying document explained that this rating meant KMS was a firm 'about which great caution should be exercised or about which little or nothing is known'.[20] The DSO staff were instructed to keep category 'C' firms 'at arms length' if contacted by them, at least until colleagues were consulted.

This rating meant such firms would miss out on marketing assistance from the DSO, but in reality this would have done little to impede a company like KMS, which was already extensively well connected. (Ironically, in 1985 KMS had actually been able to place an advert in a brochure about 'Internal Security – Counter Insurgency', which was published by a company wholly owned by the Ministry of Defence, an issue Ken Livingstone would later raise in Parliament.)

Indeed, KMS now had dozens of personnel in Sri Lanka, led by Brian Baty. His colleague Joe Forbes was a liaison officer on the Joint Operations Command, effectively the apex of the country's military decision-making mechanisms. Forbes was assisted there by a former SAS warrant officer and an ex-SAS staff sergeant, who was temporarily advising the government on 'information policy'. The company was also 'heavily involved' in psychological operations, and the British ambassador believed that recent reward schemes for information on weapons and explosives, as well as awarding medals and slicker press releases, 'all bear their mark and are reminiscent of Radfan [Yemen] or Northern Ireland policy'.[21] When the ambassador met Baty and Johnson in early December 1985, the KMS leaders told him they had three dozen staff on the island. The company was also keenly aware of the island's demography, telling the ambassador that the Sri Lankan army's latest priority was to protect the minority Sinhalese community in Trincomalee Harbour because 'too many have been leaving and the ethnic balance is in danger of changing in the Tamils favour'.[22] In fact, the company was so deeply embedded that Johnson and Baty were able to enjoy a 'private dinner party' with President Jayewardene.[23] There was 'no one else present', the British ambassador noted. 'This shows the level at which KMS operate.'

Although the envoy felt left out in the cold, the Ministry of Defence's secretive Defence Intelligence division compiled its own more detailed report on the company, which suggested it was not as disconnected from Whitehall as the DSO had assumed. 'My main conclusion', a civil servant wrote, 'is that a foreign government would have reasonable grounds on which to assume that KMS could be operating as an unofficial arm of the British government.'[24] They went on to note that KMS and its sister firm Saladin had a very high percentage of former SAS, Special Boat Service, Parachute Regiment, Royal Marines and Special Branch men involved in their organisations. Saladin had provided bodyguards for the Sultan of Oman's visits to Britain, and the company also had close links with Saudi Arabia. Saladin had received Ministry of Defence approval to submit some training proposals to the Saudi Interior Ministry the previous year and had also been 'permitted to use an Army training area'. In Sri Lanka, the official believed the company's contract was 'highly lucrative' and that KMS pilots flew Avro transport aircraft and helicopters, 'one of which is supplied with gun-pods'. In a telling conclusion, the Ministry of Defence said: 'KMS operates in a grey area'.

As 1985 drew to a close, the leader of the Tamil political party, TULF, met Foreign Office minister Lady Young in London. There, Appapillai Amirthalingam tried to impress upon her the magnitude of the suffering: 'Whole villages, particularly around Trincomalee, were being destroyed by the Security Forces', he told the minister.[25] He even claimed to have affidavits from witnesses that ex-SAS pilots were involved in air attacks on Tamil communities. By now, the minister would have been aware of a chilling statement from the head of the Foreign Office's South Asia Department, which said: 'We believe only KMS pilots are currently capable of flying armed helicopter assault operations in Sri Lanka.'[26] Although this confirmed the company's pivotal role in the war effort and corroborated what the Tamil politician was saying, diplomats decided not to radically reign in the company. Instead, defensive lines for any hostile media enquiries were duly drawn up. If asked by journalists about KMS, a Foreign Office spokesperson would solemnly say: 'We have no powers to intervene. Understand some KMS personnel are British. They are not mercenaries.'

* * *

January 1986

The New Year is often a time for reflection and resolutions. Certainly, Britain's high commissioner in Sri Lanka paused for thought about the way the war was going. Despite the British government's determined efforts since 1983 to distance itself from direct involvement in the conflict, by January 1986 this envoy was considering a radically different approach that would effectively make KMS redundant. 'I too have wondered from time to time whether we should not offer more assistance in police and military training and whether it might not be better to offer a higher degree of on-site training' – Stewart's innermost thoughts slipped out. 'The idea is attractive and could result in the Sri Lankans disengaging themselves from the KMS.'

A ban on mercenaries was not the only way to stop KMS. The British government could simply squeeze them out by doing their job itself, and probably have more control over the outcome. Stewart was mindful that the company's staff were becoming more visible in Sri Lanka, increasingly likely to take part in combat operations and therefore potentially being killed or captured. Such a scenario would provide 'a propaganda beanfeast for the militant groups and very hard words from India', Stewart remarked. There was also an economic consideration, with the company costing Colombo two or three million pounds each year, which would now be worth around eight million. Accordingly, 'The Sri Lankans would probably leap at a cheaper or indeed free alternative.' Stewart speculated that Whitehall could subsidise the cost of training Colombo's military apparatus and thereby undercut KMS, killing its lucrative contract. Perhaps Stewart was also jealous of the access the company had gained. 'Jim Johnson has become very close to President Jayewardene,' he observed, 'and appears to have a very considerable influence over him.' Stewart then commented snidely that several military commanders were 'lukewarm' on the effectiveness of KMS training and may welcome a 'more professional input' from serving British soldiers. Despite Stewart's clear preference for the British government to takeover the KMS contract in Sri Lanka, he lamented that his scheme was, alas, a 'fantasy'. Direct UK support for an 'overbearing Sinhala majority against a downtrodden Tamil minority' would be intolerable to India and the Non-Aligned Movement, who were still wary of the old colonial power.[27]

Like any prudent ambassador, Stewart kept his eye on the bigger picture. He reminded his superiors in London that the Soviets were 'constantly on the look out for cheap targets of opportunity' in countries such as Sri Lanka. In cooperation with the US, Stewart kept 'a regular eye on the Russians' and his defence attaché, Richard Holworthy, photographed any Soviet military personnel that passed through the island, studiously obtaining their passport numbers. Stewart was clear what was at stake if Soviet influence grew, singling out the island's geostrategic harbour in the north-east. 'I cannot see any likelihood of any Sri Lankan government turning to the USSR for help (and making concessions over e.g. Trincomalee) in the foreseeable future', he reassured, 'But if there were a major breakdown in the economy or in law and order, the drowning man would grasp at any straw.' In such a scenario, Stewart said, 'I hope the straw available would be a Western one'.[28]

From this lucid telegram, Britain's envoy in Colombo made it clear that Whitehall could end the KMS operation in Sri Lanka if it was willing to offer serving British personnel instead, but this would provoke an Indian backlash. As things stood, it was better to have some British military aid on hand for Sri Lanka such as that supplied by KMS, to ward off any risk of Soviet penetration in the Indian Ocean and specifically at Trincomalee Harbour.

The implications of this strategic inaction were that KMS pilots began to hit their targets harder, and more brazenly. A month later, the Sri Lankan government admitted for the first time to using air strikes on built-up areas. The High Commission explained that as they had no forward air controllers, essentially sentries on the ground who guide pilots towards their targets, 'these strikes must be indiscriminate'.[29] New Delhi was rapidly losing patience. Indian's foreign secretary, Romesh Bhandari, began describing the situation privately as 'genocide' and said the role of the British pilots was 'unhelpful'.[30] By now Whitehall went as far as accepting internally that it was 'probable' KMS pilots had 'taken charge' of helicopter attacks.[31]

The aerial bombardment was most noticeable in the Jaffna peninsula, a densely populated zone. In the space of a few days, diplomats noted that villages were repeatedly bombed by fixed-wing planes and helicopters, causing local residents to protest.[32] When the helicopters were not shooting, they were deploying soldiers into these neighbourhoods, from Jaffna in the north to Mannar in the west and Vavuniya in the centre. Britain's envoy commented that 'the government is clearly attempting to use intimidation of the civilian population'.[33] The Indian authorities made a point of telling Western diplomats that they were concerned by reports that Sri Lankan helicopters gunships 'were being flown by Brit pilots'.[34] Meanwhile, the LTTE were taking matters into their own hands. Realising that the balance of power could shift against them if the air strikes went unopposed, their spokesman, Anton Balasingham, said the group intended to acquire surface-to-air missiles such as SAM-7s within the next six months as an air defence system.[35]

British diplomats monitored the news anxiously to see if the UK media would report on KMS pilots, as any publicity could rile Delhi.[36] However, Indian diplomats were no longer as bothered about KMS as Whitehall expected. Their envoy to Colombo, Dixit, said 'HMG should not be too concerned' about India's reaction to KMS because they 'were well aware of HMG's inability to control KMS'.[37] He said that publicly the Indian government had to deplore the use of mercenaries in Sri Lanka, but that privately he accepted there was 'a large pool of ex-military personnel' in Europe and North America who wanted to 'market these skills' and if it was not KMS then it would be another 'cowboy' outfit involved.[38] He still took a dim, almost mocking, view of the company and complained about KMS pilots staying in a five-star hotel in Trincomalee and conducting daily bombing missions that did not always go to plan.[39] One KMS pilot had 'a very narrow escape' during an incident in January near Jaffna where a helicopter was grounded. An Indian official told the British envoy: 'The KMS pilot had been whisked out of the area by another helicopter and the 3 KMS pilots concerned immediately returned to Colombo saying that they were not here to be shot at.'[40] When tackled about this incident, Baty did later concede that one helicopter flown by KMS had been hit by bullets, 'although the damage was small'.[41] This near miss underlined the British consul's main concern, that a KMS pilot would be shot down, captured by rebels and subjected to a 'show trial in front of a kangaroo court'.[42]

By early April 1986, Britain's foreign secretary, Geoffrey Howe, demanded to know whether KMS pilots were involved in the bombing of Jaffna.[43] His emissaries in Colombo duly relayed the reports of civilian casualties caused by helicopters in Jaffna, but tried to downplay the company's role. 'We have no evidence of any such KMS involvement', they claimed, 'But the rumours about "white faces" flying helicopters circulate widely.'[44] A colleague described this update as 'fairly reassuring', however local Tamil civilians were growing desperate. The citizens committee in Velvettiturai said that firing from helicopters had damaged a large number of civilian houses and that over 15,000 local people were 'living in fear for their lives and basic survival'. He concluded that the residents 'are left at the mercy of "Almighty" now'.[45]

It was only at this point in the carnage that the Foreign Office's legal advisers finally began to consider suspending British passports from KMS members. This would have made it considerably more difficult for the company to recruit veterans in the UK, and severely impeded the ability of leadership figures like Walker and Johnson to fly frequently in and out of London. A Whitehall lawyer said it would be 'the only effective deterrent action we could take'. The measure was permissible in 'very rare cases' where a person's activities are 'so demonstrably undesirable that the grant or continued enjoyment of passport facilities would be contrary to the public interest'.[46] There was a precedent from 1960 when British mercenaries were warned against fighting a UN force in the Congo. However, Whitehall stopped short of taking this step. The Foreign Office also speculated that it would be hard for murder charges to be brought in Britain against KMS members who had caused deaths while serving in the forces of a friendly government.[47]

Almost instinctively, Whitehall preferred to use the old boy network rather than the criminal justice system to exert pressure on the company. The Tory peer, Sir Anthony Royle, was mobilised to have a quiet word with his best man, Colonel Johnson, as he had done the previous year. He stressed there was particular concern over KMS men bragging about their exploits at hotels in Colombo. Colonel Johnson immediately took the point, saying he would be down on his men 'like a ton of bricks' and placed one of their popular haunts, the Tangerine Hotel near the STF base at Katukurunda, 'out of bounds'.[48] This subtle intervention was typical of the chaotic response inside Whitehall as to how to handle the company, with influential diplomatic, military and political figures such as Harding, Holworthy and Royle almost always giving KMS the benefit of the doubt. For example, on a visit to London in March 1986, Holworthy had given the Foreign Office South Asia Department 'a fairly reassuring description' of the way KMS conducted itself in Sri Lanka.[49] Holworthy claimed to see 'a good deal' of Brian Baty who he said kept him 'pretty well informed of what was going on'. The defence attaché accepted at face value assurances from Baty that the KMS pilots would not fly on operational missions, and even spoke in favour of the company's work with the STF. Holworthy relayed a claim by Baty that his men were 'drumming it into' the STF commanders that they must not alienate Tamil civilians and 'the cardinal rule must be that they should not allow their forces to run amok'. Holworthy did not seem perturbed that this elementary reminder should have been required at all, over two years after KMS had begun training the police commandos.

Once again, Holworthy's praise for the company had a profound effect on Whitehall, and showed that even within large bureaucracies a handful of determined actors can have a decisive impact. Following Holworthy's positive evaluation of KMS, the Foreign Office told its ambassador in Sri Lanka it did not wish 'to lay ourselves open to charges by the company that we are seeking to damage their commercial reputation'.[50] The department said KMS training was 'likely, in the main, to be beneficial, not least in their ability to instil order and military discipline in the ranks'. Incredibly, the Foreign Office concluded: 'Any alternative to KMS could cause us even more embarrassment'.

There is also a profound parallel to these events of early 1986, on which the official files are oddly silent. The most senior KMS man in Sri Lanka, Brian Baty, was simultaneously being assisted by the British authorities in their prosecution of his would be assassin, whose murderous plotting had forced Baty to move to Colombo and adopt the alias Ken Whyte in 1984. The trial of Peter Jordan, a retired teacher, Marxist and Irish republican sympathiser, finally concluded in February 1986. It had resulted in sentencing Jordan to 14 years' imprisonment for attempting to bomb Baty's car at his home in Hereford. Had the British police Special Branch not apprehended Jordan, he would have gone on to target a long list of generals and spy chiefs. As such, the full weight of Britain's intelligence services must have been feeding evidence to the trial of Baty's nemesis in early 1986, precisely at the same time that another arm of the British state was prevaricating about whether to wash its hands of Baty and his men.

* * *

22 April 1986

Conservative councillor David Walker strolled into Elmbridge Town Hall for one last time. It had been four years since he was elected, and his tenure was up. He had decided not to stand for re-election. This was hardly surprising, given he had not showed up to a council meeting for the last five months. Still, the mayor politely thanked Walker and seven other retiring councillors for their 'many hours of time and dedication given by them for the benefit of the community'. The remainder of the council joined in wishing their departing colleagues the best of health and good fortune.[51] They had no idea what Walker would do next. He was no longer a serving councillor, and could once again focus entirely on his business.

A few weeks after Walker resigned as a councillor, KMS pilot Tim Smith settled down for a briefing with Brian Baty at the company's tranquil office in Colombo. It was a converted suburban house somewhere in the capital, possibly on Ward Place, near to the upmarket area of Cinnamon Gardens. The building was kitted out with electronic security features, storm shutters and mercifully, air conditioning. Baty proceeded to bring Smith up to speed on the company's work in Sri Lanka, with the police, army and now the air force. There was one nasty surprise though. Baty told him blandly that the British High Commission 'have washed their hands of us'.[52] Smith was shocked, but Baty was probably trying to reinforce the 'deniable' nature of their contract, as by now the white pilots were beginning to spark some media interest in the UK. In reality, Baty was still in regular contact with embassy staff such as the defence attaché, Richard Holworthy, who staunchly supported the company's work.

About a week after Smith landed in Sri Lanka, questions were tabled in Parliament. The Labour MP for Bradford West, Max Madden, asked the foreign secretary: 'What representations he has received about Britons flying helicopter gunships against Tamils in Sri Lanka'? Probing further, the left-wing politician demanded to know 'what instructions have been given to the British High Commission in Sri Lanka in relation to British citizens who are involved in military action against the Tamils'?

Madden's questions were fielded by an unflappable junior Foreign Office minister, Timothy Eggar MP. At just 34, Eggar was a rising star of the Conservative Party. Educated at Cambridge, qualified as a barrister and with a career in investment banking already under his belt, Eggar entered Parliament in the 1979 general election, which saw Thatcher come to power. An early proponent of privatisation, Eggar was an apt choice to deal with enquiries about British mercenaries. In his written answer, Eggar assured Parliament that:

> We have received no formal representations about Britons flying helicopter gunships in Sri Lanka nor about Keeni Meeni Services (KMS). We understand that KMS has for a number of years provided training to the Sri Lankan security forces. The details of the agreement under which such training is provided is a commercial matter between the company and the Sri Lankan Government. We have, however, made it clear to both that we do not wish British subjects to become involved in the fighting. The Sri Lankan Government have assured us that KMS personnel would not be employed in an operational role.[53]

Whitehall had just played its trump card, flourishing President Jayewardene's 'assurances' that it had kept tucked under its sleeve for almost a year. What Parliament was not told was that Britain's high commissioner never had any confidence in those assurances, and that Whitehall knew perfectly well that KMS pilots were carrying out combat missions.

Away from this stray parliamentary question, Tim Smith's transition from sleepy Carlisle to Sri Lanka started smoothly with a few relaxing days sight-seeing in Colombo, meandering through the crowded side-streets with an attractive German woman he had sat next to on the flight over. Meanwhile, his wife was struggling to deliver his child thousands of miles away in Carlisle. Smith spent some time talking to her on the phone, while he waited for his new colleagues to return from a mission in the north of Sri Lanka. The army and air force were trying to advance on Jaffna from a base at Elephant Pass, the narrow sand spit connecting the ancient Tamil city to the rest of the island. The operation was timed to tip the balance of power in favour of the Sinhalese forces just before another round of peace talks got underway. However, 22 mines in the first six kilometres stopped this master plan dead in its tracks. Demoralised young soldiers broke down in tears at the roadside and refused to go a step further, Smith learned from his old pal and fellow KMS pilot Greg Streater, who appeared to have given air cover to the Sinhalese troops as they retreated.[54]

Ten days after he arrived in Sri Lanka, Captain Smith was finally put to work. The squadron leader announced that Smith had three days to learn how to fly the Bell Huey 212 helicopter, before being dispatched to Kankesanturai, a port suburb on the northernmost coast of the island, where he would be able to appreciate the conditions facing the air force.[55] Smith says he was 'astounded' at being sent so far from Colombo and straight into a combat zone. Kankesanturai was traditionally a Tamil pilgrimage site, which had been taken over by the Sri Lankan military at the start of the conflict and converted into a high-security zone. The local Tamil residents were unceremoniously turfed out and not compensated for the loss of 6,000 acres. It was compulsory purchase, Sri Lankan style, and an old colonial airstrip at Palaly was rapidly put into use for military flights. Smith dived into his crash course on the helicopter, and collected an assault rifle, 120 bullets and a flak jacket. Fearing what was to come next, he packed 200 cigarettes, some books and canned food for the flight north the next morning. He convinced himself he would only be an observer in the north for a few days before returning south to train young pilots.

When he touched down in Palaly, probably on 28 May 1986, the base was surrounded by sandbagged walls, under a virtual state of siege. The only other white man on the flight had been a young journalist from *Solider of Fortune* magazine, who Smith assiduously avoided. His home on the base would turn out to be a breeze-block extension to an existing red-roofed bungalow where the Sinhalese pilots lived. The 'whitefaces' as they were known, had a corrugated tin roof at their end of the building. A row of flowerbeds rimmed the gravel drive, beyond which were watchtowers and blast walls, constructed from sand, earth and rocks to neuter incoming mortars. It looked like a rustic retirement home ready for the apocalypse.

Once he had settled in, a young flying officer, Namal Fernando, gave Captain Smith a briefing. The squadron's main task at that time was to support and resupply all the army outstations that were cut off by the LTTE, from Mullaittivu on the north-eastern beaches, to the jungle town of Kilinochchi further south, across to Karainagar naval base to the west of Jaffna city. 'We would carry coconuts, cabbages, chillies, cartridges, conscripts and constables, cannons, casualties, correspondents and cameramen', Smith noted with a poetic glumness.[56] He was expecting to be there for a few days. Fernando informed him it would be more like a few weeks. And the food was worse than the accommodation. Breakfast consisted of fried eggs on toast. Lunch was lukewarm curried cabbage, swimming in oil. He cheered up momentarily when another KMS pilot descended from a helicopter and sauntered over to his bungalow. It was Dave Wharton, one of the company's Rhodesian employees, undoubtedly a veteran of the country's unsuccessful bush wars against black African nationalists, before relocating to apartheid South Africa. Perhaps this was the pilot whose gruesome war stories of door gunners slaughtering children had tipped Robin Horsfall over the edge a few weeks earlier. But Wharton had kept working for KMS, and seemed to avoid giving Smith the same harrowing insights on his first night in Jaffna.

Smith would be left to learn for himself what his job truly entailed. On his third day at the remote airbase, he was scrambled to an incident at Mandaitivu, one of the small islands on the archipelago south-west of Jaffna. Smith was in the left seat, effectively co-pilot, with an intercom wired into his helmet so he could speak to the control tower and crew. The Bell helicopter was well armed with forward-firing weapon pods, which the pilot, Flying Officer Namal Fernando, controlled from the cockpit. Sinhalese crew in the cabin behind them operated the sideways-mounted door guns, which consisted of a 0.50 inch heavy machine gun plus 7.62 mm side guns. Soldiers down at the Mandaitivu outpost had seen some boats on the lagoon, and fearing the Tigers were laying a trap, called in air support. All Smith could see though, from 1,500 feet up in the air, was an old man on a bicycle, pedalling slowly. Fernando was taking no chances, and called for the weapons to be armed. Smith flicked the circuit breaker on the control panel, effectively removing the safety catch from the pod gun. Fernando rolled the helicopter into a shallow attack dive and locked the weapons sight.

'What have you seen?' Smith asked naively. 'I think he is maybe a bad guy', Fernando answered childishly. 'If I shoot at him and he runs away, it will prove that he is.' The old man did not run away. The stream of bullets knocked him from his bike and he 'lay very still'.[57] Smith could not bring himself in his memoirs to explicitly say he had just helped to kill an old man. Before he had a chance to process this, Fernando was chasing after two motorbikes that had aroused the suspicion of the army. The door gunners promptly opened fire, until they too fell from their bikes and 'lay still', which seems to be Smith's euphemism for death. 'A hollow, empty feeling filled my stomach', he said, belatedly re-engaging the safety catch on the weapons system. 'I should have done that 20 minutes before', Smith wrote mournfully. 'Perhaps the harmless old man on his push bike would still have been alive.' Captain Smith was now an accomplice to what almost certainly amounted to the cold-blooded murder of a civilian, and had seen at first hand what Horsfall had only heard as hearsay. Local priests confirm this incident took place, noting at the time: 'Road transport to the Islands is suspended due to army camps being set up at both ends of the causeway. Large number of people wading through the lagoon to cross over to Kayts strafed by a Helicopter. One person is killed and another injured.'[58]

Still, Smith continued to work for KMS, a job that routinely involved facilitating atrocities against the Tamil civilian population. 'I discovered that anything that moves following an attack was "in season"', he rued.[59] By 5 June 1985, after only a week flying with Namal Fernando, Smith says he had 'watched' at least five people die. To take his mind off these deaths, he went on three-mile runs along the coastal edge of the airbase every evening. For safety reasons, he took his assault rifle. Later at night, with his KMS colleagues, conversation took a darker turn. They muttered about the morality of what the Sri Lankan pilots were doing, and how the company had hoodwinked them into being caught up in combat missions. 'Flying gunships had never been mentioned when we had signed our contracts for the trip', Smith recalled. In fact, the company had specifically denied any suggestion of it. The assembled mercenaries also pondered why the ground troops never followed up on their aerial attacks, to find out who had been killed or wounded, and perhaps bring a casualty in for questioning to obtain information.[60]

Their introspective laments were interrupted by nightly bombardments of their camp from LTTE mortars and machine-gun fire. It was a baptism of fire for Smith, but the company's pilots had worked in this combat zone for some time. Local priests had already recorded over 50 attacks by helicopters between August 1985 and Smith's arrival in May 1986, throughout which KMS pilots had played an instrumental role in keeping the aircraft flying.[61] These attacks had involved methods of warfare as sinister as: the throwing of acid from a helicopter; the shooting dead of a one-year-old child cradled in its mother's arms, who herself lost an eye; the burning of houses; the destruction of farms and livestock; the death of a baker and a priest; the fleeing of hundreds as refugees; the disruption of education for thousands of students; and the firing on first responders as they tried to put a corpse into a van.

*   *   *

On the evening of 6 June 1986, Smith's contract was radically revised. Kapila Ratnasekera, one of the Squadron's flying instructors, casually asked Smith if he was willing to fly the gunship as the captain, and not just the co-pilot. This was a step too far for Smith, who refused. He would remain in his role of 'safety pilot', to stop the helicopter getting shot down if Fernando did anything too risky. Smith knew this was a 'fine distinction', and that Fernando was a 'bloodthirsty, murderous bastard'. But he needed the money, and in his words put 'money before morality'. Looking back, Smith proclaims that if the air force had asked him to captain the gunship 'as a means of reducing the chances of opening fire on innocent bystanders' then he probably would have done it.[62] It seems as if this notion – that the KMS pilots were somehow a moderating influence on the slaughter, and not an integral component vital to keeping the planes in the sky – is a fiction that many involved sought comfort in.

At it turned out, the killing continued – wherever Smith sat in the helicopter. Shortly before lunch on 7 June 1986, Smith and Fernando were mobilised to locate a green bus just outside the perimeter fence. It was believed to be carrying, among other passengers, several gunmen who had shot at the main gate to the airbase, before deciding that a local bus was their most suitable get-away vehicle. Smith's helicopter quickly chased it down as it headed towards the small town of Achchuveli, and the door gunner sprayed it with bullets. When the bus swerved off the road, Smith thought his crew would leave it for the army to move in and take over. Instead, his door gunner continued to fire as the passengers – men, women and children – fled from the bus. Eventually, Smith shouted 'Bloody hell, Namal, there are people in those houses, the bloody mainstreet is full of innocent people, you can't keep on with this.'[63] Namal Fernando reluctantly gave the order to ceasefire.

Assuming armed Tamil men had indeed boarded the bus and used it as some kind of 'human shield' for their escape, then the Tamil militants had committed a war crime. But by opening fire on the bus, knowing it had civilians on board as well as armed men, the Sinhalese helicopter crew were also guilty of a grave violation, in which local sources suggest two people were killed and another two injured.[64] Smith tried to pre-empt any such charges in his memoirs, by claiming that he intervened and halted the incident. While this is superficially appealing, it does not entirely absolve him of responsibility since he helped the helicopter take off and locate the target, knowing his colleagues were 'murderous' and had no qualms about killing civilians.

* * *

Nicaragua, May 1986

After a career spanning over two and half decades, Colonel Robert Dutton had finally retired from the United States Air Force. He was a decorated Vietnam war pilot and part of a special operations vanguard that had been trusted with some of the White House's most clandestine and questionable adventures, from Laos to Iran. This gave Dutton a formidable network of social contacts that would ensure him lucrative employment in the next stage of his life. He was close to a former air force general, Richard Secord, a fellow special operations veteran who was now running Stanford Technology Trading Group International, also known as the 'Enterprise'. This was one of the main shell companies in the Iran–Contra scandal. When Dutton went on a trip to Colorado with the Secord family, he mentioned that he was planning to retire in the summer of 1986. General Secord then made him an offer to 'come and work for him' at the Enterprise. 'At that time he mentioned that he was in the process of conducting a special operation that he could sure use my help at', Dutton later recalled.[65]

When Dutton duly retired from the military on 1 May 1986, stepping down as director of operations at a US air base, he commenced the same role for General Secord's company the very next day. 'It's what I'd been trained to do', he said of his new role directing airfield operations, which had a monthly salary of $5,000. For the next six months, Dutton would be working on a special operation to resupply the Contras in the south and north of Nicaragua by air. 'I was asked to manage the operation, to basically take an operation that was not working and get it to work.' For this job, Dutton reported to General Secord and Colonel Oliver North. Although Dutton says he was not at that time aware of North's role in the White House's National Security Council, he sensed that North had 'very broad contact ... at the very highest levels of the various branches of the government', including with the CIA director. Crucially, North indicated to Dutton that 'we were working for the President of the United States'.

US policy on Nicaragua in the summer of 1986 was particularly hawkish. Senior officials estimated that there were 10,000 Contras who were 'an untrained peasant army' and would require significant congressional funds to turn them into a 'viable guerrilla force'. A British diplomat said: 'The US intention is that the Contras should become capable of making life sufficiently bloody – literally and figuratively – to create a real groundswell of popular opposition to the regime.' The UK official thought, 'they are probably over-optimistic. But the Americans clearly intend that the situation in Nicaragua should get much nastier before it gets better.'[66]

Dutton was a hands-on kind of guy, who didn't like sitting around in Washington. Reports suggest that Dutton was based out of Ilopango Airport in El Salvador, with a team of around 19 personnel on the base, from air crew for flying the missions to 'maintenance folks' who 'took care of our small fleet of aircraft ... they were the kinds of guys that could put together an operating aircraft with bailing wire and chewing gum', he recalled.[67] On a visit to Central America in mid-May 1986, Dutton found that the resupply efforts were poorly organised, and he would have to spend the summer hampered by the rainy season trying to make it work.

'I understood that the equipment was not operating very well, that we were not getting aircraft in the air', Dutton complained. 'There were problems in making contact with the forces we were supposed to be resupplying. I wanted to go down and see, for myself, exactly how the operation was being run – and who was being effective and who wasn't.' Dutton had four aircraft at his disposal: a C-123 plane, with about a 2,000 mile range and 10,000 pound payload; and two smaller C-7s, with about half the range and payload but still capable of a round trip to southern Nicaragua. They also kept a single-engine propeller-driven Maule, which was 'sort of our taxi when we needed to get out to the forward operating bases rather than flying one of the large aircraft'. The main operating base in El Salvador had a formidable warehouse replete with parachute rigging, uniforms, light machine guns, rifles, ammunition, mortars, grenades and C-4 plastic explosive.

In addition to this fearsome arsenal, Dutton found some other assets on his first trip to the main base in El Salvador. There were three Britons on the base, who Dutton initially mistook for South Africans. 'The initial plan had been that we would hire – or that the Brits would be hired – to fly the missions that would have to actually go inside Nicaragua', Dutton explained. 'We did not want to have to expose Americans to those kind of flights. So we had hired two pilots and a load-master to fly those missions.' The plan was hatched and outlined to him by General Secord and Oliver North, who had paid none other than David Walker for the British crew. KMS would have been familiar with working inside El Salvador, where it had provided bodyguards at the British embassy not so long ago. Secret documents show that this contract would be far more lucrative than their work for Whitehall. Walker was to be paid at least £110,000 for his pilots (now worth over £300,000), which was among several 'big ticket items' for North's operation which would 'nearly wipe us out'.[68] The money was paid through Gouldens, a prestigious City of London solicitors firm that Walker used.[69] The UK Foreign Office would later comment privately that 'Washington sourced allegations that KMS employees had been working in El Salvador helping arrange resupply air drops to the contras in Nicaragua may be true'.[70]

Although Walker was clearly held in high regard by Oliver North, Dutton was far less impressed at the quality of his pilots, who he says ultimately never flew into Nicaragua despite the original plan. 'We thought we were flying highly experienced airlift pilots. And it turned out, the one with the most flying time was a helicopter pilot, and the other one did not have the experience that we required. And they just – they didn't work out operationally, and then they had some problems with the local general, and were asked to leave', Dutton said disparagingly. By mid-June, Colonel North authorised missions by American-manned crews inside of Nicaragua. 'We've got a southern force that is desperately in need of support. The idea of using the British crews has not worked out. We'll fly our own missions', he reportedly told Dutton.

While Dutton was in El Salvador meeting the KMS pilots, others were trying to revive an old plan to help the Contras. On the evening of 21 May 1986, Margaret Thatcher's personal secretary, Charles Powell, made an unexpected phone call to a staffer in the Foreign Office's Central America department. Thatcher's aide asked if the diplomats had seen any sign the US might be trying to procure Blowpipe surface-to-air missiles from Chile, 'presumably for ultimate use by the Contras'. Powell did not say what had prompted his enquiry, although there had been some recent press and parliamentary interest about the Contras acquiring the weapon. The diplomat denied hearing any suggestions about this, but began to look into the matter.

It transpired that the Chilean Air Force attaché in London had gone to the Ministry of Defence earlier that week for an 'off the record' chat. He had asked the DSO whether it would be willing to lift the end-user certificate on the Chilean army's 50 Blowpipe missiles that the UK had supplied to the country. This appears to have been a repeat of David Walker's December 1984 attempt to direct the Chilean stockpile of British Blowpipes to the Contras. The Ministry of Defence told the Chilean pilot that the end-user licence could not be waived. Whitehall was sticking to its guns, literally, and would not deviate from the position it had taken on the earlier Short Bros application to sell the missile to Honduras.[71]

Although the Foreign Office had no evidence the US had prompted Chile to make this latest approach, the department felt 'there must be circumstantial grounds for suspicion', given the previous attempt to acquire Blowpipes for the Contras.[72] Colonel North was certainly still determined to get hold of surface-to-air missiles. On 10 June 1986, a few weeks after the Chilean pilot in London sheepishly approached the Ministry of Defence, Colonel North sent a message to President Reagan's national security adviser to take stock of the various efforts to secure arms for the Contras from third countries. North still hoped Thatcher would come to the aid of the Contras. 'We should look to going back to Maggie [Thatcher] on the Blowpipes if we are going to do anything at all about outside support in the next few days', North said. 'I would love to carry the letter from RR [Ronald Reagan] to her if we are going to move on something.'[73] While this document shows Colonel North was persistent in trying to find a way to persuade Whitehall to supply Blowpipes, there is no evidence that such missiles ever reached Nicaragua, and ultimately the Contras managed to find alternative models of such weaponry from other countries. The UK government insists that it 'at no time authorised the supply of arms to the Contras',[74] however, it was clearly discussed at some point, although not at a prime ministerial level. A Whitehall official wrote furtively: 'We are not in a position to deny that no approach was made to HMG by the US Government about the supply or diversion of Blowpipe [...] Neither can we deny that the subject was ever discussed between a British official and Oliver North.'[75]

9

Grenades in Wine Glasses

Sri Lanka, June 1986

Despite the humiliating rejection of KMS pilots in Nicaragua, the company's aviators remained in hot demand in Sri Lanka. The only issue was Smith's refusal to captain one of the gunships – a small act of insubordination that had not gone unnoticed by the KMS high command. Sometime between 7 and 10 June 1986, both Brian Baty and David Walker arrived at Kankesan-turai, 'resplendent in tropical lightweights, like a couple of white planters'.[1] These two most senior KMS commanders proceeded to berate Captain Smith and his colleague Dave Wharton, accusing them of disloyalty to the company and 'lacking in moral fibre'. This language inevitably riled Smith, who felt they had tricked him into taking on a combat role rather than being a flying instructor. He did his best to ignore their criticism. 'Those two people didn't have a clue anyway', he wrote furiously. 'They just sat in their air-conditioned offices, hundreds or thousands of miles from the action and counted the cash.'[2] Meanwhile, Smith felt that he and the other KMS pilots were involved with killing Tamils on a daily basis.[3]

The Tamils also felt the same way. One of them, the general secretary of TULF, Amirthalingam, tried to raise it over dinner with a British diplomat in Madras, south India. The Tamil politician said a district magistrate had been ferried to a post-mortem in a helicopter flown by a British pilot, and that 'KMS were flying supplies to military camps'. The UK diplomat rebuffed his concerns, claiming that there was a 'distinction' between 'flying on operational missions and merely flying supplies or personnel around'. When Amirthalingam persisted, and pointed out that 'there was plenty of evidence that "white pilots" flew helicopter missions', the diplomat demanded that he produce detailed evidence to show KMS was involved'. How exactly Tamil civilians were supposed to obtain such evidence, while under fire from the skies, was of course not specified.[4]

KMS pilots were not only stationed on the north coast at Kankesanturai, or at the international airport near Colombo. By mid-June 1986, they were also based at air strips in Vavuniya, as well as in the Eastern Province, which was under the control of the STF. The main cities there, Batticaloa and Trincomalee, had KMS pilots flying in and out of them.[5] The Tamil militants, for their part, were not oblivious to the fact that white men were increasingly in the cockpits of the helicopters that caused them so much damage. They monitored the military's radio channels, and Smith observed an increase in radio chatter whenever he or his KMS colleagues spoke over the airwaves. If they had managed to down one of the KMS pilots, it would have marked a major victory for their armed struggle, and constituted a doomsday scenario for the nervous British diplomats.

Fortunately for the latter, Captain Smith was skilled at giving covering firing to Sinhalese soldiers, often halting the Tamil militants in their tracks. On 20 June, he took over the controls from Flight Lieutenant Roshan Goonetileke, and talked the door gunners through laying down air cover. With Smith in charge, his helicopter delivered 1,000 rounds of ammunition onto the Tamil militants beneath them at Tondamanna, on the northern coastline. He killed twelve Tigers that day, and showed Goonetileke how he had manoeuvred the aircraft in order to sustain such a withering assault. Goonetileke would go on to become an air chief marshal, and the commander of the SLAF from 2006 to 2009, when the country's worst ever war crimes were committed. Later, he became chief of defence staff and then a presidential adviser. Such is KMS's long legacy in Sri Lanka.

Captain Smith continued to fly out of Kankesanturai throughout June, July and August 1986, with only one weekend off in Colombo. During these three months, his helicopter played a significant role, on a daily basis, of suppressing the Tamil uprising against a seriously demoralised and inexperienced army of Sinhalese conscripts – around 5,000 of whom had already deserted.[6] Without the air cover and fast casualty evacuations, it seems likely that the army would have been routed that summer. Whitehall was not completely oblivious to the major role being played by British pilots over the summer of 1986. On 16 July, a day before Smith's thousand-bullet sortie, Lieutenant Colonel Richard Holworthy returned from a visit to military units in the Eastern Province. There, he spoke to a KMS helicopter pilot who described his role in repelling an attack on Mankerni, an army outpost along the coast north of Batticaloa, which included strafing militants with rockets and machine-gun fire.[7] A local Sri Lankan colonel, who was listening to the conversation between Holworthy and the KMS pilot, joked about 'white mercenaries at Mankerni … elderly mercenaries!'[8]

The British embassy regarded this incident as 'the first clear evidence to reach this mission of direct KMS involvement in combat flying'. The diplomats realised that the answer given to Parliament in May was now on tenuous factual footing. It was the foreign secretary himself, Geoffrey Howe, who appeared most concerned by this cable and wanted to ramp up the pressure on KMS to avoid combat operations.[9] Over the next few days, more details of the incident at Mankerni emerged. The KMS pilot involved was an ex-British army air corps pilot like Tim Smith. The mercenary said he was scrambled in a night-time emergency and flew as co-pilot, insisting he took no active part in the shooting.[10] The company's supporters in Whitehall seized on this comment to downplay the incident and suggested the only cause for concern would be if the pilot 'repeats his account before other, less discreet listeners'. Some diplomats favoured giving KMS a mild slap on the wrist – another oral warning from

Harding to Colonel Johnson, this time delivered in London to avoid jeopardising Holworthy's 'valuable' information-sharing relationship with Baty that existed in Colombo.

Once again, it was Howe, the foreign secretary, who had to keep pushing back against the KMS sympathisers within Whitehall. He warned diplomats they were relying on a 'very fine distinction' between being a co-pilot and a pilot in helicopters that were flying combat missions.[11] The Tory minister also appeared somewhat concerned about misleading the House of Commons, and did not feel ministers could 'in all conscience' continue using the same line to Parliament, which relied on President Jayewardene's assurance that British pilots would not fly on operations. Howe wanted to become 'much tougher on KMS' and was unconvinced Whitehall had fully considered 'all the possible levers'. He asked wishfully whether the UK had any anti-mercenary legislation that was 'used in the 1970s to prevent various organisations enlisting men to serve in Africa?' Unfortunately for Howe, Whitehall's deliberate inaction after Diplock's report into the Angola mercenary scandal, partly to protect companies like KMS, was now coming back to haunt the department with a vengeance.

Howe's concerns found some sympathy with a Foreign Office lawyer, who felt her previous legal advice on KMS had been misinterpreted. She pointed again to the possibility of withdrawing passports from the company's staff and echoed Howe's opinion that the distinction between pilot and co-pilot was 'without a difference so far as the real issue is concerned'.[12] Britain's foreign secretary and a government lawyer had come out strongly against the company, but instead of following their lead, diplomats remained willing to bat for KMS. A British embassy official argued that if the KMS helicopter had not intervened at Mankerni, 'the army camp could have been at least badly mauled, at most, overrun'.[13] The debate over KMS was by now a cross-Whitehall issue. The Department of Trade and Industry (DTI) was tasked with investigating the company's structure,

223

effectively producing a third report to supplement the earlier two compiled by parts of the Ministry of Defence. The DTI found that all known directors had apparently taken up formal residence in the Cayman Islands, as well as moving their registered address to the Channel Islands in March 1985. As such, KMS no longer had a place of business in the UK.[14] This deliberate shift to two of the world's most secretive tax havens thwarted the DTI from investigating any further.

In this carefully constructed vacuum of information, some diplomats tried to downplay the extent of KMS links to the Foreign Office, claiming that they had only ever employed the company to train local guards at 'one or two of our posts in the Middle East' in the mid-1970s. However, more astute colleagues recalled a 'potted history' of the company whereby it had in fact been hired as recently as 1982 during the Falklands conflict. 'We always had a very good service from KMS, both as regards the quality of their men and the speed of their response', one diplomat breezed. The Beirut contract in particular had proved rather more durable than other colleagues assumed, spanning 1976 to 1982. (One official even noted that after the KMS team finished guarding British diplomats in Kampala, they were hired by the European Commission after a verbal recommendation from the Foreign Office.)[15]

And yet as much as some diplomats, and even the foreign secretary, may have wanted to distance their department from the company, whenever the situation reached a crisis it was almost always the KMS sympathiser, Deputy Under Secretary Sir William Harding, who was assigned the task of walking away from the firm – resulting in a predictable fudge. On 8 August, a day before Colonel Johnson returned to Colombo for the rest of the month, he met with Harding in London. It was their fourth meeting to discuss the company's Sri Lanka contract. As always, the colonel was bullish and unrepentant. Yes, the KMS staff training the ground forces were in 'daily contact' with Brian Baty. But the pilots were a different matter altogether. They were on secondment from the company to the SLAF and 'might not be seen for weeks'. Incredibly, the colonel claimed: 'KMS were not allowed or encouraged to see them [their pilots]; there was no day-to-day contact'. This hands-off arrangement sounds unlikely, given that Smith was visited by Baty and Walker at his northernmost airbase. But the colonel insisted the SLAF directed the KMS pilots and the company was not consulted about their tasking.[16]

Harding then realised that if KMS had no direct control of its pilots then logically it could not be certain what they were doing. The colonel pushed back and asserted his men only ferried supplies and troops into Jaffna, although there was a risk of coming under fire while leaving the airfield because the runway extended into the jungle and was not sufficiently guarded by Sri Lankan troops. Eventually Johnson conceded his men were flying helicopter gunships and 'sometimes intervened in ground actions'. Perhaps more importantly, the colonel confirmed that KMS were responsible for recruiting, paying and administering the pilots even while they were with the SLAF.

When Harding brought up the precedent of removing passports from British mercenaries fighting in the Congo 25 years earlier, Colonel Johnson batted it away as irrelevant. He then tried to play something of a trump card, claiming that Margaret Thatcher had 'mentioned that KMS's activities were acceptable to her' when she met President Jayewardene in October 1985 at a Commonwealth summit in The Bahamas. This oral assurance had apparently been followed up with an exchange of letters, to confirm the prime minister's approval in writing. This revelation rather took the wind out of Harding's already lacklustre sails, and Colonel Johnson pushed his advantage home. He emphasised how the British High Commission team themselves accepted KMS had 'raised the Special Task Force from a scratch force to a body of men with a fine record which had pacified the area south of Batticaloa and which was the envy of the Sri Lankan army to the extent that KMS were now also retraining the Army a company at a time'.

In a similar vein, KMS pilots would raise the standards of local airmen, who he said tended to 'blast every village in sight'. He insisted: 'Somebody had to take on the job' and the SLAF were incapable of air-to-ground missions 'unless a white face was present'. By this point, the two men appeared to agree to disagree, with Harding still hoping KMS would find an alternative to using British pilots. The pair, as always, parted amicably, and Harding naturally did not seize the colonel's passport to prevent him from flying out to Sri Lanka as planned the very next day. Instead, he looked forward to a report upon the colonel's return after the August bank holiday. Another opportunity to stop KMS had been missed, with devastating consequences for civilians living under the company's flight path. The company's influence on this stage of the war cannot be understated. Sri Lanka's national security minister was of the opinion that 'without KMS assistance to the armed forces the terrorists would take over'.[17]

By now Whitehall was beginning to accept that KMS pilots would fly on operations in Sri Lanka for much of the next twelve months. When Colonel Johnson returned from Sri Lanka in early September, he helped defuse the situation by promising the Foreign Office he would reduce the number of pilots 'currently on an operational basis' from six to four as soon as possible, a move diplomats could spin as a small victory.[18] Although the department had been unable to locate the convenient letters of approval between Thatcher and Jayewardene, Whitehall no longer seemed too worried about whether or not such papers existed. Harding thought a small reduction in pilot numbers was 'probably as good as we were going to get and there was no need at present to take up Colonel Johnson's offer to come in to repeat it to us'. Further relief came when it emerged that Sri Lanka's defence secretary, General Attygalle, would spend five days in London over mid-September to 'agree with KMS a reduction in the number of British pilots employed'. Although the company had moved its registered office to the Channel Islands, and domiciled their directors in the Cayman Islands, they were clearly still keen to use London for important meetings with one of their most valued clients. And the British government had decided, once again, that it would do almost nothing to stop them.[19]

* * *

For the KMS pilots, the reality of their work was a world removed from the calm telegrams between London and Colombo. In early September 1986, just as the company was promising to scale back the role of British pilots, Captain Smith was dispatched to Sri Lanka's war-torn east coast. He was stationed there with the 'murderous' Namal Fernando in China Bay, Trincomalee Harbour, where Sinhalese troops were searching small towns in Koddiyar Bay, on the southern rim of this vast port. Civilians were fleeing the soldiers as Smith hovered above with Fernando's finger in control of the trigger. 'I should have known that the trip would be a real bastard', the Englishman reflected. The noise of the helicopter circling low overhead terrified the civilians, who dived for cover in the jungle. All apart from a grey-haired and skinny old man 'clothed in flapping rags who, in total and utter desperation, climbed a tree'. Smith recalled what happened next: 'When we left the area there was a tattered rag-covered body hanging out of that tree.' Fernando had killed him, in an act that Smith said sickened him and would haunt him for decades to come. He tried to persuade Fernando to cease fire, which just made him angrier and select another old man, this time on a bicycle, for the side gunners to slaughter. Fernando finally ordered the side gunners to cease fire. 'We left the old man there, a lifeless heap on the road', Smith lamented, regarding Fernando as a 'depraved little animal'.[20]

While Smith is keen to adopt the moral high ground in this incident, and many others, it is worth analysing his own culpability in such events. Within days of starting to fly helicopters for

KMS, he knew that his Sinhalese colleagues willingly carried out war crimes, and faced no repercussions. They also had impunity to conduct more atrocities. Even if Smith initially believed that he could moderate their behaviour, he must have realised that after three months this was wishful thinking. By continuing to co-pilot the aircraft, he was a knowing accomplice to the war crimes. A further, damning factor, is that immediately after the cold-blooded murder of these two elderly Tamil gentleman, a bullet pierced the Smith/Fernando helicopter during their trip back to China Bay to refuel. The bullet landed in the side of Fernando's armoured seat, sending ceramic chips into his eyes and a plume of fibreglass dust into the cockpit. Captain Smith quickly took over the controls and flew the helicopter to safety, preventing it from crashing and allowing Fernando to live to kill another day. British diplomats quickly gleaned remarkably accurate details of this incident, confirming that the Sri Lankan pilot was injured and the KMS co-pilot managed to land the helicopter safely.[21] The high commissioner noted: 'The incident is of course a sharp reminder of the risk of KMS personnel being killed or captured.' For Smith, it was a danger he hardly needed reminding about.

* * *

When Robin Horsfall gave me his reasons for quitting KMS, he explained that there were no journalists on the ground who he could brief about the atrocities, and thus no way of shaming Colombo into reigning in its forces. It is significant then that Smith was presented with a chance to alert international media but chose not to. A British journalist appeared at Smith's helicopter camp shortly after the murders of the old men in China Bay. The reporter had befriended the camp's commander, but Smith refused to speak to him. This was a missed opportunity to highlight the war crimes that had taken place – effectively ensuring that any wrongdoing by his unit was covered up.

Furthermore, although Smith is at pains in his memoirs to portray himself as a highly professional flying instructor, the reality is that most of the men he was working alongside were amateurs. They were young, Sinhalese conscripts who did not want to be in Jaffna, living under a state of siege in isolated outposts, hundreds of miles from home. Some of the soldiers Smith evacuated had shot themselves in order to be taken off the battlefield. Again, without the KMS pilots, it is highly likely the Sri Lankan army would have had to abandon the north, effectively giving the Tamils a *de facto* state of their own.

This scenario was not as far-fetched as it might sound. In 1986, the Sri Lankan army was becoming so demoralised that the Tigers, especially in their strongholds like Prabhakaran's home town of Velvettiturai (known locally as VVT), were able to establish permanently manned sentry towers around key buildings, ringed with concrete walls and trenches. They were now so powerful that it was no longer necessary to hide as a guerrilla army – instead they could fight in the open like a conventional force. This meant they could attack the army's outposts with sustained machine-gun and mortar fire whenever a helicopter tried to resupply them, putting Smith's job at immense risk. By mid-September 1986, Smith was refusing to land at Point Pedro or VVT outstations, and instead the KMS pilots would only give air cover while the Sinhalese pilots took their helicopters in to land. 'Two Whitefaces on top cover; shades of Western mercenaries commanding gunships', was how Smith characterised the arrangement.

The situation began to alarm the local army commander, who could not understand why the KMS pilots were refusing to resupply his besieged troops. On 20 September 1986, the military's chief of staff personally visited Kankesanturai to confront Smith and his KMS colleagues. It was virtually a court martial situation, with the mercenaries forced to justify their insubordination. Smith took the lead in explaining the dangers that they were facing because the army was so demoralised and refused to shoot at the Tigers when the helicopters tried to land. He success-fully swayed the chief of staff to their point of view, to such an extent that the commander asked Smith to advise the army how to clear out the Tiger positions in VVT. 'We must give the Tigers a bloody nose, we must whack those bloody buggers. You must co-operate with these pilots to sort them out', the military chief told the army.[22] Again, Smith had made a decisive intervention.

That night, the army closed in on the LTTE stronghold, killing two and capturing three. The next day, Smith was allowed to operate his helicopter how he wanted, and his crew fired 7,000 bullets at the LTTE position, to help cover the army on the ground. Smith says he does not know how many they killed, only that 'They kept coming, on foot, on pushbikes and in cars. We knocked them down and they crawled on until we hit them again.' After this epic battle, the Anglo-Sinhalese helicopters were safe to fly in and out of Point Pedro once more. The protest by the KMS pilots had led to them giving tactical advice to the head of Sri Lanka's military about how to suppress a rebel base near to the LTTE leader's home town, such was the impact the company's men were having on the conflict. And yet the Tigers' support network on the Jaffna peninsula was so deeply rooted that it took them less than a fortnight to recover and commence counter-attacks.

For the Sinhalese conscripts, this fresh onslaught was a drain on their already low morale. For Smith, he was counting down the days until he went on leave in England, thousands of miles away from this war in which he was playing a pivotal role. He seemed fairly certain that he would be back for another tour though, despite everything he had seen and done, and the way KMS had tricked him. A pack of dogs scavenging near the edge of the airfield brought his mind sharply back to the present, as he went on an evening jog. The animals were fighting each other around a smouldering fire that Smith was using as a landmark on his run. Worried they might attack, he lobbed some small stones to ward them off. One dog snarled as an object fell from his mouth. It was a human bone, with charred flesh still hanging off it. It slowly dawned on Smith that these were most likely the remains of four captured Tamils who he had seen arrive at the camp a few days earlier, before they suddenly disappeared. They would have needed helicopters to transport the prisoners to internment camps in the south, but Smith had never noticed them leaving. It appeared that the prisoners had been killed and their bodies burned. This was exactly the type of horror that had spurred Horsfall to quit KMS. Captain Smith, despite witnessing the results with his own eyes, would react very differently.

When it was time for his first tour to end, he flew down to Colombo expecting to find Brian Baty at the company's office in the capital's suburbs, only to learn he was elsewhere on the island. The most senior KMS man on the island clearly did not prioritise debriefing one of his key aviators after a tumultuous five months, in which Smith estimated he had been 'personally involved in the death of 152 Tigers' before ceasing to count. With Baty otherwise engaged, Smith had to make do with Wing Commander Oliver Ranasinghe, the Sinhalese air force officer who had first briefed him at Katunayaka. The English pilot spent the first ten minutes grumbling about coming under fire when landing at army outposts, where the soldiers provided no cover. He also complained about being expected to captain the gunship, when he had only signed up to be an instructor. The Sinhalese officer responded with flattery, saying he admired Smith's work and was 'extremely happy with your performance'. Neither of the men, however, appeared to raise the issue of repeated atrocities, from the machine gunning of the old man who climbed the tree in Trincomalee to the charred remains of prisoners in Jaffna. What Smith wanted was for a senior colleague to appreciate macabrely what he had done, warts and all. By the end of the debrief, Ranasinghe was confident Smith would come back for another tour. As the men shook hands, Ranasinghe had one last request: 'I don't suppose you could bring me back a dozen of those wine glasses that you get from the petrol stations in the UK when you come back?'[23]

* * *

November 1986

Smith's one-month leave in England did not go well. He drank heavily to numb flashbacks from the front line, as he struggled to cope with undiagnosed post-traumatic stress disorder. He kept seeing a Tamil woman in a blue-green sari who walked slowly away from her husband's tractor after Smith's crew had machine gunned it to pieces on Elephant Pass. 'I was filled with an unrecognised self-disgust. I wanted a home life full of tranquillity that I had so denied to so many Tamils.' Smith's solution was to make it worse. On Tuesday 15 November 1986, he strolled through Colombo airport and caught a lift back to the base. The armoury reissued his assault rifle and 120 bullets. He did not bother collecting a flak jacket for this tour. Then he waited patiently for orders. An SLAF camp seemed more normal to him than his family home in Carlisle.

Not much had changed, although he was stationed slightly further from the front line at Vavuniya, on the island's central spine. The retrained army commando unit also had a presence at this camp, and took advantage of Smith's flying skills to practice covert helicopter insertion into the jungles.[24] The so-called Trackers were a welcome change from the timid conscripts in the camps across Jaffna, crying, self-harming and constantly needing a resupply of coconuts. These men were much more like the elite SAS units Smith knew from Northern Ireland, who spent days fending for themselves in the fields of South Armagh. Smith was thrilled with this increased professionalism, and relished the opportunity to work with them. When the Trackers became embroiled in a gun battle with Tamil militants at an elephant sanctuary, Smith flew in to extract them from the jungle and rescue a casualty. As he scurried into the hastily arranged landing site, Smith coordinated with the special forces soldiers on the ground, told his gunners where to lay down suppressing fire and marvelled as the Trackers streamed into his cabin and immediately resumed shooting into the jungle to cover the take off. Although Smith apparently despised Baty, he conceded that the KMS stalwart had produced an effective fighting force.[25]

Not everything was in favour of the mercenaries and their companions however. Smith's close friend from Carlisle, Greg Streater, suffered a stroke to his left side one night trying to rescue an army patrol from the jungles near China Bay, Trincomalee. His helicopter came under heavy fire and it seemed to send him over the edge. The KMS man was evacuated back to the UK. Next, Smith's old flying partner, Namal Fernando, was hit by a rocket-propelled grenade in Mannar. The safety pin stopped the rocket exploding, but it seriously damaged the helicopter, suggesting that the Tigers were beginning to master basic air defence skills.

In response, Smith took on an increasingly aggressive role. By this stage he had formed the opinion that to 'come to grips with the Tiger a lot of innocent civilians would have to die'. On one sortie, he said: 'Bugger the village, blow 'em away, literally.'[26] So when an army patrol on the Mannar peninsula was pinned down by a group of Tigers firing from a schoolhouse in early December, Smith scrambled the gunship and flew as the primary pilot. 'The dividing line between morality and necessity had grown dim', Smith noted unflinchingly. He proceeded to personally fire the podguns into the school building, hoping there were no kids left inside. 'Bloody good shooting', he later remarked. Mercifully for any children sheltering inside the school, all the helicopter's weapons suddenly froze. Captain Smith's attack fizzled out.

I tried to trace Smith to talk to him further about this incident, but to no avail. He was last seen sailing into the sunset in Spain, and he did not respond to messages I left him on a veterans' version of Facebook. However, Britain's former defence attaché,

Richard Holworthy, appeared to relish discussing the subject with me, as we idled among the acres of rolling orchard adjacent to his Bordeaux barn. When we ventured back inside, I passed Holworthy something I had found at the National Archives. It was a report he had written several months after Smith's attack on the school, in which he said: 'There is no doubt at all that the KMS pilots have been of great assistance in keeping standards up, and whatever the political pros and cons of the KMS flying operations are the effect on the air force – and the army that they support – is beneficial'. Holworthy went further, and said 'I, for my part, am happy to see them here (and relieved to be flown by them)'. Despite the British government claiming it disapproved of KMS pilots, it was clearly content for its embassy staff to use them like taxi drivers whenever the diplomats needed to reach remote parts of the country.

I then raised Smith's account of the school shooting with Holworthy, and asked how he reconciled that with his praise for the company in this report. His response took me aback. 'You've read his book have you?' He smiled disarmingly, evidently familiar with Smith's memoirs. 'But those are war crimes', I suggested to him. 'Those are war crimes, yes', Holworthy agreed nonchalantly. He then gave a more graphic example of what the KMS pilots had done. 'Initially actually they were talking about bombing from helicopters', he began ominously. They took First World War-era Mills grenades, 'pulled the pin out and put the grenades into wine glasses, with the wine glasses holding the lever', Holworthy gesticulated. 'Then they'd fly over and drop the grenade with the wine glass, and of course when it hit the ground the glass broke, the grenade exploded', Holworthy laughed. 'I know this was happening because when I went to Trincomalee one day and had lunch there and there were no wine glasses at all – the whole bloody lot had been dropped on Tamils with grenades inside them!'

Perhaps this explained why the SLAF officer had asked Smith to buy him wine glasses from the petrol station in England –

to replenish their arsenal. 'I thought it was a slightly dodgy operation', Holworthy said with classic British understatement. 'It was the Sri Lankan themselves who told me this, you know quite proudly', he added. By now he was casually describing their atrocities: 'The Air Force was strafing using rockets and eventually bombing, fairly blanket bombing. They had no worry as to where they were dropping them – civilian casualties just didn't come into it, I think because the only civilians were Tamils. And they didn't really worry about that.'

However, the civilians were not only Tamil. On 25 September 1986, a group of engineers for the international German broadcaster, *Deutsche Welle*, were travelling towards their radio masts in Trincomalee. The area was prone to fierce fighting between the militants and government forces. Suddenly something fell from the sky and killed one of the German media workers, Ulrich Heberling. His colleagues were 'in no doubt' that he was killed by 'a grenade from a helicopter'.[27] Heberling left behind a wife and child.[28] In 2012, one of his colleagues, Dieter Pelz, commented: 'I remember Ulrich well for his cool character and good humour. His death was clearly avoidable.'[29] At the time though, British diplomats had a rather more narrow focus. One said: 'Our main concern has of course been to find out whether a Sri Lankan Air Force helicopter, possibly with a KMS pilot aboard, caused the German's death. We cannot yet rule out this possibility.'[30] In fact, it seems from the available archival evidence that they were never able to rule it out.

* * *

Washington, December 1986

Shirley Napier had worked at the Enterprise for just over three years, on and off. It had been a mixture of full-time and part-time employment, as she juggled her secretarial work with studying. The job started well. Setting up an office from scratch, sorting out the files, doing the accounts, answering the phone. Her boss, Richard, or General Secord, was a decorated US military man and she respected him. He recognised her potential, promoting her to staff assistant and taking her on foreign work trips. She was a trusted part of the Enterprise, officially known as Stanford Technology Trading Group International.

One of the perks were her occasional visits to the Old Executive Office in Washington, DC, on Pennsylvania Avenue, next to the White House. She would scurry across its chess-board-pattern plaza and dive inside to deliver envelopes, cash, a Bible and encryption machines. Her boss was doing substantial business with an important man called Oliver North, and Napier would hand over this eclectic mix of items to Fawn Hall, North's secretary. Sometimes she ran errands that seemed a bit strange, but she tried not to dwell on it. Once a manager asked her to collect $16,000 dollars from a man in Miami. She flew down there and found him as soon as she walked out of the arrivals gate. He showed her the money, stuffed inside a large FedEx overnight envelope. Mrs Napier took the package into the ladies toilet to count the cash before boarding the next flight back to Washington, where she gave the envelope to North's secretary. Then she went home, and decided it was best not to tell her husband what had just happened.

In mid-March 1986, Napier had travelled with General Secord to London, where he met privately with David Walker. Napier did not know who he was at the time, but after this trip Walker visited the Enterprise's office in the US twice. 'Mr Secord had told me that he was the English version of our Delta Force', Napier said of Walker's business, comparing KMS to the US army's special forces unit.[31] Walker was clearly someone significant, because shortly before Napier went on her Christmas vacation in December 1986, General Secord made her spend several days shredding documents that he had personally selected for destruction. These included a series of Rolodex cards for three men: Oliver North, Cuban dissident Rafael Quintero, and David

Walker.[32] General Secord was bracing the Enterprise for the day when their cover would be blown – and he wanted to ensure David Walker's name stayed out of it.

Oliver North, however, was not being so diligent. Rather than shred his most sensitive documents, he hoarded them in his office safe, expecting that to remain a sanctuary. In fact, it proved to be a hostage to fortune. Within months, US congressional investigators would gain a warrant to crack open his strong box and uncover a perplexing diagram. In scruffy handwriting, North had drawn a cryptic map of the clandestine Contra resupply operations, linking General Secord's shadowy bank accounts to arms dealers. Amid the soup of acronyms, one stood out: KMS.

* * *

Kokkadicholai, Eastern Sri Lanka, January 1987

KMS and its supporters claim that the company's training improved the discipline and professionalism of Sri Lanka's armed forces. While this might have been true of the small Tracker team, the same cannot be said of the police STF. By January 1987, after receiving three years of KMS training, its ranks had swelled to encompass 1,000 elite officers – but it was more murderous than ever before. Half an hour's drive south of Batticaloa, across a lagoon and away from the crowded coastal settlements, lies a quiet patchwork of paddy fields nestling around the village of Kokkadicholai. In the 1980s, it was home to a thriving prawn farm. Today, the village is notable for its red and yellow pyramid-shaped monument bearing the names of 85 people who died on 27 January 1987. Over three decades later, the relatives of the dead still gather at this stone statue every anniversary to mourn their loss.

The killing started early in the morning. An armada of helicopters thudded towards the prawn farm. It is not known whether Tim Smith was personally involved in the operation, although the company's aviators were flying over the Eastern Province that month.[33] When the helicopters arrived at the prawn farm, they delivered a mixture of army and STF personnel. 'We suddenly heard a lot of noises outside, while we were sleeping at 5am', one survivor told me. Pakkiyaselvam Ariyanethiran would later become an elected MP for this area, but that morning he had to flee for his life. 'Everyone was screaming that the army was coming and that we had to run', he said. 'I was only wearing a shirt and a sarong. I lost my sarong when we were hiding in the field so I had to run in my underwear.' Ariyanethiran was one of the lucky ones who escaped. Anyone who was caught in their path was executed, including teenage boys. Troops threw the bodies into a local well, permanently polluting the village's precious water source. It is regarded as one of the worst massacres by the security forces from that entire phase of the war.

The massacre had widespread ramifications, disturbing even hardened observers who would normally give the STF the benefit of the doubt. Lieutenant Colonel Holworthy reported that the task force's image had been 'tarnished by the massacre at Kokkadicholai'. He took an unusually critical tone, noting how the police commandos were 'a 100% Sinhalese force' which was hated by the Tamils. 'I think it will be some time before the east can [be] said to be pacified', he remarked.[34] However, many local people had simply had enough of waiting to be 'pacified'. Those who could fled the country, with some opting for the former colonial power. They arrived to a hostile reception – the most unfortunate found themselves locked up at Harmondsworth detention centre near Heathrow Airport. Thatcher's government was turning the screw on refugees, and Tamils were bearing the brunt of it, but the Home Office had underestimated their resolve. After years of war, the Tamils were by now a highly politicised community who knew how to agitate for their rights.

As deportations loomed, organisers like Varaimuttu Varada-kumar from the Tamil Information Centre rallied the refugees to resist, in the most media-savvy way they could think of. When a coach-load of 60 Tamils was driven to Heathrow Airport against their will, the men stood on the runway and removed their trousers. Footage of their protest made the evening news, and Varadakumar took to the airwaves, denouncing the prawn farm massacre on the BBC. The community's actions won a brief respite from the deportations, as Britain's immigration minister, David Waddington MP, flew to Sri Lanka on a 'fact-finding mission' to establish whether the country was as safe as the Home Office imagined. Waddington met Tamil professionals on his trip who him told him that the STF continued to conduct 'mass roundups of terrorist suspects and was given to terrifying displays of military might'. Waddington nevertheless concluded that it was safe to deport Tamils to Sri Lanka and he believed more KMS training was needed, suggesting that any assistance KMS could give to further training Sri Lankan forces 'would be welcome'.

The Home Office was determined not to show any sympathy whatsoever towards the Tamil refugees. Until 1987, many of the arrivals were detained in Britain's nascent network of immigration detention centres, from Ashford in Kent to Harmondsworth near Heathrow. Often they were put in remand centres, alongside people facing trial for serious crimes, for weeks on end. Some were lucky enough to have a backbench MP, like Jeremy Corbyn, raise their case with the home secretary and avert their deportation (in those days MPs had such powers).[35] But even when Tamils were released, they faced racist violence in the communities where they were housed, like Newham in east London. A firebomb through a letterbox on Burgess Road killed three Tamil refugees in November 1986. The East End was not always safer than the Eastern Province. Flames haunted the Tamil people wherever they went.

Meanwhile, the Home Office had devised a new way to torment Tamil people: water. In January 1987, as the massacre raged at the prawn farm, a British civil servant proposed that a ferry should be converted into a floating detention centre. His idea was met with approval from David Waddington's superior, Home Secretary Douglas Hurd, who said such a vessel was 'unlikely to appear too luxurious or too spartan'.[36] By May 1987, Tamil refugees were forced to board the *Earl William*, a roll-on roll-off ferry now permanently moored in Harwich harbour. Among the shipmates were some of the Tamils who had escaped deportation in January by stripping off on the Heathrow runway.

∗   ∗   ∗

A week after the slaughter at the prawn farm, Colonel Johnson and David Walker strode into the Foreign Office in London for a frank hour-long meeting. Their sympathetic contact, Sir William Harding, had now been replaced by another man, David Gillmore, but he still seemed to give the company a lot of credit. Harding's replacement did not mention the massacre at Kokkadicholai, and instead told the pair 'we had no problem with the training of ground forces'.[37] However, ministers still felt 'extreme anxiety' at the presence of British pilots on operational flights and these had to stop, a view 'endorsed in full by all concerned'. At this point the renegade colonel resurrected an old canard, the pledge of support given to President Jayewardene by Thatcher when they met in 1985, and subsequently cemented in writing, by letter. As such, he professed to be 'mystified' by the British government's latest request to desist. He went on to deny that KMS were 'egging the Sri Lankans on to a military solution', nor did they intervene in policy making or try to tell the government how to run the war, a claim which seems to directly contradict earlier statements he had made about the company's influence.

Indeed, Johnson does not seem to have been a man of his word. Contrary to previous promises to phase out the British pilots, he said the Sri Lankans had insisted 'the less efficient non-British pilots' should be the ones to go, meaning that the remaining team of KMS aviators was exclusively British. The KMS leaders refused to rule out their pilots being involved in operational areas, although they claimed to have reduced the number of KMS pilots from eleven to eight. Lady Young, the Foreign Office minister, was unimpressed by the colonel's obstinacy, and urged her department to put their concerns to KMS in writing, a move supported by the UK envoys in Colombo and New Delhi, who felt such correspondence would be useful to placate both the Indian government and British Parliament should a KMS pilot be shot down.

Meanwhile, the company was creating new facts on the ground, effectively running rings around any attempts by Whitehall to reign it in. Their army training team had moved from the relatively peaceful camp at Maduru Oya to Palaly air base in Jaffna, which was under constant attack from Tamil militants. This KMS team was half a dozen strong, and led by another colonel, John Gordon-Taylor, who appears to have been a battle-hardened Dhufar veteran.[38] The idea that a KMS team this near the front line was purely training troops and had no involvement in directing operations borders on the unbelievable. The company now had 38 staff scattered across Sri Lanka. The largest contingent, 17 men, were still training the STF, despite the prawn farm massacre. The pilots had vacated Palaly, but they were near enough to fly there when necessary. A KMS sniper instructor was training the army commando regiment, a company man remained embedded with the Joint Operations Command (despite Johnson's claim not to intervene in policymaking) and Baty continued to be stationed in Colombo with a quartermaster. The company found room to expand its remit further, and started training naval land units, ensuring KMS was influencing all three branches of the Sri Lankan military, which was finally beginning to encircle the Tamil militants.

Colonel Johnson had also bought himself some time in Whitehall by raising the canard of the letter between Jayewardene and Thatcher. By this point, allegations were beginning to swirl that Thatcher had supported KMS operations in Nicaragua, so diplomats had to seriously consider the prospect that the prime minister had covertly given her blessing to the company's work in Sri Lanka. Civil servants carefully tried to 'get a feel for the atmosphere' in Downing Street before they probed too far.[39] The pressure on KMS was ratcheted up another notch when Nick Davies, one of the best investigative journalists of his generation, ran a story alleging that 60 KMS staff in Sri Lanka 'walked out after complaining that the government troops they trained have committed a series of atrocities against the minority population'. The Foreign Office seemed surprised at how negatively the article portrayed the STF, and expected it to trigger a cabinet meeting the next day.[40]

The reaction among embassy staff in Colombo on the other hand was more muted. Britain's defence attaché, the KMS sympathiser Richard Holworthy, reached out to his trusted contact, Brian Baty, who predictably denied the report had any substance.[41] He rubbished the idea of a mass walkout by his men, conceding that only a handful of staff had quit. 'Two of the temporary replacements for those KMS members who were casevaced [casualty evacuated] to the UK last year having refused to serve in Willpattu (a game park in Mannar district where militants operate) were returned to the United Kingdom', but it is not clear if Baty was referring to STF instructors, army trainers or the pilots. (Baty had previously told Holworthy that six KMS men who were training the STF had returned to Britain after contracting a form of hepatitis.)[42] He did acknowledge that there could be some unrest in his ranks, saying that some KMS personnel were 'reluctant to serve at a training camp in Palaly', which was hardly surprising given that the 'training camp' was in reality a front-line military base on the Jaffna peninsula.

The envoy then began to put a gloss on the human rights situation, claiming that 'by all accounts' the STF had behaved in a 'reasonably well disciplined manner' until January, a statement that suggested the copious warnings from Tamil civil society and even the CIA had been ignored by British diplomats. He did acknowledge that some of the members of the STF were 'undoubtedly responsible for the murder of 20 plus innocent civilians' at the prawn farm, although the actual death toll was four times that amount. Despite this, the envoy claimed that the security forces' human rights record had improved during the past six months. Stewart signed off by sticking the boot into the journalist's source, speculating that the allegations could have been made up by the two KMS men who refused to serve in a dangerous part of the company in order to justify their stance. He noted that one of the two men who Baty said had quit, Curry, was 'now referred to by his colleagues as "chicken curry"'.[43]

However, others in Whitehall were more discerning of KMS. The Foreign Office had spoken informally to Charles Powell, Thatcher's top Downing Street aide, who assured them that no letter between the prime minister and President Jayewardene endorsing KMS activities in Sri Lanka existed.[44] The KMS sceptics in Whitehall now prepared to make their move by drafting a written warning to the company's chairman. Three years since KMS commenced offering its services in Sri Lanka, and countless massacres later, the British government would finally formally register its displeasure at just one aspect of the firm's full spectrum support for Colombo – the pilots in operational areas.

The high commissioner was shown the draft, in anticipation that the KMS high command might call on him before it could be dispatched. Sure enough, Johnson and Walker dropped by the embassy on 7 March, where Stewart was flanked by his deputy and Holworthy. The KMS chairman was clearly on edge, admitting that US congressional investigators had recently found a handwritten note with the company's name on it buried inside Colonel North's safe, as they probed the Contra operations. He described the 'KMS London office in Kensington High Street' as being 'under siege by TV and press', which Walker and Johnson were not looking forward to encountering when they returned there at the weekend. Finally the renegade colonel's confidence was beginning to fade and he appeared 'more anxious about the fuss he would be facing on return to London than about KMS activities on the ground here'.[45]

There would certainly be one letter waiting for them upon arrival. It was addressed to Colonel Johnson at 7 Abingdon Road, the terraced office in South Kensington where both Horsfall and Smith had been recruited by KMS. Gillmore had taken a step that his predecessor Harding would never dream of, putting into writing the foreign secretary's concerns about KMS pilots on operational missions in Sri Lanka. Yet, the criticism remained restrained, noting that the British government 'cannot prevent KMS from entering the employment of a foreign government, nor would we wish to discourage you from continuing to train the armed forces of Sri Lanka'.[46] It took a fortnight for the company to reply, but eventually David Walker did send a written response to the Foreign Office in which he offered to brief the department on the outcome of their future trips to Sri Lanka. The letter was sent from 7 Abingdon Road, thereby acknowledging that KMS did indeed have a place of business in the UK despite its utilisation of tax havens elsewhere.[47]

The mood in Whitehall was now much more relaxed – having fixated on grounding the pilots, they now felt the company had caved in to their demands by allegedly removing the last of its aviators from Palaly on 6 March. The fact that diplomats continued to believe that KMS personnel 'may be involved back stage in the operations room in Colombo' was not really a concern, although in reality it carried far-reaching ethical and legal implications for the mercenaries involved. What mattered to Whitehall was that they were unlikely to be captured by Tamil rebels and spark a diplomatic incident. In a rare moment of self-reflection, a civil servant admitted: 'It has to be said that our attitude towards KMS training has been ambivalent over the years.' He also alluded to Thatcher's support for the mercenaries, reasoning that 'KMS are much valued by President Jayewardene and the Prime Minister has made it clear to us that she wishes us to support him in his efforts to restore law and order in Sri Lanka'.[48]

In reality, KMS had barely scaled back its aviators. When Walker returned to Sri Lanka in late March, he found that one of the company's pilots remained at the front-line Palaly airbase. However, the difference was that the Foreign Office felt it had secured a degree of influence over the company, and Walker had become 'relaxed and friendly' in his interactions with the department.[49] When Walker 'implied' that the KMS pilot was withdrawn from Palaly once he spotted him, the Foreign Office appeared delighted by the changed attitude of KMS. 'Well done!' One diplomat scribbled on top of a telegram. 'It seems we may be getting somewhere ... It might be a good idea to debrief Walker after his next trip, provided we can do so without appearing to be in cahoots.' All it took for the company to neutralise the British government's concern was a minor redistribution of its aviators. Every other aspect of the KMS's footprint in Sri Lanka, its work with the STF, army, snipers, naval guards, logistics and operational planning, could remain intact, simply because those liaisons were kept behind a barbed wire fence and would not be noticed by the public should they go wrong. With the company's tentacles enmeshed across all of the Sri Lankan security forces, the Tamil militants would finally begin to feel the squeeze.

10

Bugger Off My Land!

As David Walker's military operations in Sri Lanka reached their gruesome climax in mid-1987, the intricate web of plausible deniability he had spun to protect his mercenaries across the world began to come crashing down. KMS was about to face sustained pressure in the most dangerous arena it had ever faced: the United States Congress. On 20 May, a former Coca-Cola executive sat down in front of a media circus in Washington, DC. The floor of Congress was full of photographers, all pointing their cameras and flash bulbs at this balding man with white hair, large spectacles and a strong Central American accent. All available seats were crammed with reporters, anxiously awaiting his every word. Adolfo Calero, the Contra leader himself, was about to testify.[1] Oliver North's secret war in Nicaragua began to implode once Congress realised that its ban on funds for the Contras was being circumvented by the White House. Finding KMS named on a document in Colonel North's safe was just the start. By mid-May, key players in the scandal had agreed to testify in return for favourable treatment. Calero, 55, was a friend of President Reagan and accustomed to stomping around Washington's corridors of power drumming up support for his cause.[2]

Now he was about to reveal what some of his more militant activities had entailed. The Nicaraguan government's helicopters had always posed a lethal threat to Calero's guerrillas, and KMS had initially hoped to destroy the choppers on the ground before they were airborne. To further this plan, an associate of

Walker made a reconnaissance mission to Nicaragua, where he gathered intelligence on the security surrounding the military airbases. However, the KMS associate returned in a pessimistic mood. 'The access to those helicopters on the ground was difficult', Calero claims Walker's team had concluded. 'It didn't go beyond that.' A sceptical senator asked the Contra leader to clarify whether he discussed hiring Walker to destroy the helicopters on the ground. Calero calmly responded: 'Well, that was the idea of his taking a trip there to see what the possibilities were in making such an effort.' The guerrilla leader went on to say: 'We didn't discuss the situation any further because it was something considered to be extremely difficult.' The senator shot back: 'Difficult to destroy them on the ground?' Calero: 'Difficult to carry out the operation on account of the Sandinista surveillance.'[3]

Calero's cross-examination soon moved on to other matters, but Walker's support for the Contras was no longer secret. The British SAS veteran had organised the covert reconnaissance of Nicaraguan military airbases, with a view to sabotaging government helicopters, violating numerous Nicaraguan laws. The next day, US senators would return to the Englishman's role in this once covert war. The Democrat's most senior congressman, Thomas Foley, began to grill General Singlaub, a founding member of the CIA and avid anti-communist.[4] Singlaub had supplied arms for the Iran–Contra scandal, but Foley also wanted to ask him about other matters: 'You also testified you discussed with Colonel North the possibility of foreign nationals conducting operations, military operations, for the Nicaraguan assistance inside Nicaragua, and that Colonel North stated that he had already hired a former SAS officer. Is that correct?' The general concurred, and acknowledged that David Walker was the officer in question. Oliver North had told Singlaub 'in general terms' what KMS planned to do in Nicaragua. Although North did not divulge the 'specifics of clandestine entry and some of those details', he told the general that Walker's operation was to 'infiltrate a team of non-US personnel into the vicinity of the airfield at Managua, with a view of destroying the HIND-D helicopters that were known to be located there'. This was the same mission that Calero had revealed, although it never went ahead because of the risks involved.[5] If it had been successful, then it would have eliminated the helicopter gunships that Singlaub regarded as 'the most effective people-killing machine in the world'.

The idea that such formidable aircraft could effectively turn the balance of a war through its effectiveness as a 'people-killing machine' was something the Tamils were learning first hand as the war in Sri Lanka reached a gruesome new low. The irony of attempting to ground Nicaraguan helicopter gunships while simultaneously using such helicopters to crush the Tamil uprising would not have been lost on Walker and the KMS hierarchy. Two years of KMS flying an equally deadly design of helicopter had significantly eroded the Tamil resistance, who were now cornered in the north of the island. Days after Singlaub's testimony, the Sri Lankan authorities launched a massive offensive to recapture the entirety of Jaffna, a mission provocatively named Operation Liberation. Planes dropped leaflets urging Tamil civilians to shelter in a Hindu temple at Thikkam to avoid the fighting. But there was no escape. Barrel bombs full of napalm were dropped on the temple from Avro aircraft – planes that KMS had helped the Sri Lankan pilots learn how to fly, even if British pilots were not directly involved in this attack. One mother, whose child had lost half their face, said she saw 'a huge fireball' containing a red and blue plastic rubber substance that stuck to her child's head and had to be removed in hospital.[6] As the carnage unfolded, the KMS pilots enjoyed a temporary holiday in hotels at the other end of the country. Sensing a major operation, Whitehall had pressured KMS to ensure none of its men were directly involved, even if the company had spent years developing the Sri Lankan military's capability to mount such a full-scale attack.[7]

* * *

It would be another two months before Congress delivered the death knell for KMS. On 13 July 1987, Oliver North prepared to testify before the congressional hearing on Iran–Contra. Sitting poised in his pristine US marine corps uniform, the White House's disgraced national security aide did his best to protect his British friend from a barrage of precision-guided questions launched at him by leading Democrat congressman Thomas Foley.[8]

North admitted having 'some contact with David Walker', a man he described obliquely as 'a British subject who runs a business in the Channel Islands'. When Foley, who had originally trained as a lawyer, asked for more detail and suggested that Walker was really an 'international specialist in insurgency and military matters', North again tried to protect Walker. 'The firm that he represents or runs specializes in a number of activities like security assistance and the like, but I'm not entirely certain that we really want to go too far in this discussion in a public session.' As he delivered this understatement, North calmly conferred with his attorney, perched close to him on his left flank.

Foley, confounded by North's evasiveness, began to flounder in his cross-examination, giving time for the marine colonel to play his trump card. 'I don't feel particularly comfortable discussing it in open session. I think there are equities that belong to other governments that are at stake here', said North, alluding cryptically to the UK. The chairman sensed something was at stake and weighed in, suspending any further questioning on this topic until he had checked what could be discussed in front of the television cameras.

North had won Walker a brief reprieve, but only for five minutes. The chairman soon announced that the White House had declassified the 'David Walker material', referring to a series of top secret memos that North had written after his first meeting with the KMS director back in the heady days of December 1984, when anything – from Blowpipes to sabotage – seemed possible. Congressman Foley had used the extra time to refine his line of questioning. He now went for the jugular: 'David Walker is an international arms and security specialist, and is a British subject. Is that right?' 'That is my understanding, yes sir', North hissed back. Foley zeroed in: 'Did you authorize, or have any discussions with him regarding activities inside Nicaragua?'

North: I did.
Foley: Did you authorise him to perform military actions in Nicaragua?
North: I did.
Foley: What were those actions?
North: David Walker was involved – his organisation, as I understand it – in support of the Nicaraguan resistance, with internal operations in Managua and elsewhere, in an effort to improve the perception that the Nicaraguan resistance could operate anywhere that it so desired.

North was now throwing his British friend directly under the bus. He could hardly be bothered to camouflage Walker's sabotage of the hospital in Managua in March 1985, which gave the appearance that Contras were at the city gates and was carefully timed to take place before the crucial congressional vote on whether to restore aid to the Contras. But Foley was not finished yet. The declassified material on David Walker made extensive reference to Blowpipes. Foley asked: 'We're talking about military actions, attacks on military aircraft, of the Sandinista government?'

'Yes.'
'Did you directly authorize that?
'I don't directly – I didn't directly authorize anything. I encouraged Mr. Walker to be in touch with the people who could benefit from the expertise that he had.'

In other words, North had told Walker to contact the Contra leader Adolfo Calero to discuss the acquisition of surface-to-air missiles. 'Did you report that recommendation to Admiral Poindexter?' Foley probed further, trying to connect the scheme with President Reagan's national security adviser.

> 'I did.'
> 'Did he approve of it?'
> 'I reported what I thought was a good idea for certain expertise that the Nicaraguan resistance did not have, as a consequence of insufficient training, insufficient operational capability ... '
> 'Did he ever tell you he approved that?'
> 'He never told me not to.'

North had betrayed Walker, but he was still willing to shield one man. 'Do you know if the president was ever informed of this?' Foley asked. 'I do not know, sir', came North's reply.

The very next day in Washington, Colonel North faced another formidable inquisitor – chairman of the House Intelligence Committee, Congressman Louis Stokes. He asked Colonel North very specifically about the hospital attack. 'On March 6, 1985, an arms depot in Managua, Nicaragua, was destroyed by an explosion', Stokes said. 'What was your role in arranging for the explosion?' Colonel North was on the ropes but kept trying to defend himself from the blows. 'I personally had no role in it whatsoever. It is my understanding that foreign operatives were engaged in that activity and assisted there in.' North eventually managed to bring a halt to the questioning, by saying: 'We talked of that in the executive session the other night sir.'[9]

The 'executive session' had been held in private, away from the television cameras. But there was a transcript, which British journalists from the show *World in Action* managed to obtain. It showed that North was asked 'Was it David Walker?' – concerning the identity of the hospital bomber. North is said to have replied: 'It's my understanding that David Walker provided two technicians involved in that.' Back in the public session, where the colonel's lawyer was frantically whispering into his ear, Stokes had had enough beating around the bush. 'Who paid Walker for conducting that operation?' he seared on to the congressional record. Colonel North said: 'It is my understanding that he was paid by the Nicaraguan resistance or by General Secord.' In other words, Walker's money had come from North's right-hand man in the Iran–Contra scandal, the head of the Enterprise himself, the general who had told his secretary to shred Walker's Rolodex card before Christmas. The revelations in Washington made David Walker a person of interest in the congressional hearings. Congressmen offered him immunity from prosecution if he would speak to them and a congressional investigator spent several months in the summer of 1987 repeatedly trying to phone Walker in London, offering to meet him at a place of his choice in virtually any part of the world. Walker simply did not reply.[10]

* * *

Although David Walker's livelihood was now under threat, Robin Horsfall was already coming to terms with life as a pariah since quitting KMS. 'When I got back to Britain I was very much ostracised from the market, from the circuit, for a while, and I found it very difficult to get a job', Horsfall told me. 'So I essentially created a job of my own, running a small door team in Hereford until things picked up.' Being a nightclub bouncer in an English country town was a far cry from his glory days in the SAS. How long could Horsfall handle such a mundane line of work? 'Then, a job came down the grapevine to those of us that were living in Hereford saying that there's work available training people in Tamil Nadu', the southernmost Indian state. The Tamils were attempting to secure their own team of SAS veterans. 'That meant we were training Tamil Tigers to go and fight the Sinhalese Sri Lankan army', Horsfall explained. This job would have completed Horsfall's political transformation, from a counter-terrorism expert to a fledgling third-world freedom fighter.

The British establishment, however, had other ideas about Horsfall's job offer. 'Before I got the chance to consider it seriously enough, I had a light tap on the shoulder from some anonymous person to say we wouldn't approve of that, and if you do that, we can't stop you however we'll make sure you never work again.' The threat worked, and Horsfall declined the offer. This incident intrigued me. 'Who do you think that person was operating on behalf of?' I asked.

'He was operating on behalf of the Foreign Office.'
'And why did they not want British mercenaries doing that?'
'They didn't want British involvement on both sides of the conflict, that was the explanation I was given.'

This might sound like a humanitarian rationale, a way to de-escalate the conflict and avoid arming both sides, but I suspected there was a more sinister reason behind it. I asked Horsfall if this was another indication that the Foreign Office was backing KMS. 'Yes I think they were', he responded. When it came to it, Whitehall was willing to use threats and intimidation to ensure a particular British mercenary company had a monopoly on Sri Lanka's war.

* * *

The White Sultan of Downing Street

The Iran–Contra scandal was not confined to Congress. On the other side of the Atlantic, news of David Walker's involvement in Nicaragua had triggered questions in Parliament. Although Thatcher's government refused to concede that there should be an inquiry into KMS, behind the scenes her staff were becoming increasingly rattled by the barrage of questions from MPs. The Sandinistas were highly effective at raising awareness of their struggle among socialist politicians around the world. Their movement was a *cause célèbre* in left-wing Labour circles. Backbencher Jeremy Corbyn worked closely with the Nicaragua Solidarity Campaign, as did the more mainstream George Foulkes, who was a spokesperson for the Labour Party on foreign affairs. Foulkes repeatedly asked a series of parliamentary questions about what contact Margaret Thatcher had with David Walker, which she dodged. There was a rumour circulating that the two were personal friends, and the fact that Walker was a Conservative councillor for Elmbridge from 1982 to 1986 lent credence to this theory.

As the scandal reached its peak, a new British high commissioner arrived in Colombo. David Gladstone, the descendent of a former prime minister, was a very old-school diplomat. Born in the British Raj, he 'fell' into diplomacy for want of something else to do with his life. When I visited him in 2018, he resided at a Home Counties mansion, where Tony Blair lived in a pavilion on the edge of his grounds. I sat for hours inside Gladstone's freezing grand ballroom, surrounded by towering glass-panelled bookcases, endless mahogany tables and thick pile rugs, as he fondly recalled his career with the Foreign Office. However, when I asked him who was running KMS, he became evasive and answered cryptically: 'They had some sort of political cover in this country but I'm still not quite sure what. And I thought it was something by that stage that I didn't need to know about and therefore I wasn't going to beaver around too industriously to find out.' I asked, 'What do you mean by political cover?' He appeared to become anxious. 'I mean, I was told, on what authority I'm not quite sure, there were one or two British politicians who were connected to the company – but that is all hearsay.' Among these politicians would have been Sir Anthony Royle, who was Colonel Johnson's best man and vice-chair of the Conservative Party until 1983.

Gladstone's answer left open the possibility that there was another powerful politician connected to the company, but he claimed to be at a loss as to their identity. Other civil servants in Whitehall were also mystified about the power behind the company even as they attempted to avert a scandal. Foreign Office staff endeavoured to craft responses for the prime minister to use in Parliament, to ward off the claim that David Walker had held meetings with her. In a memo, one official wrote: 'I have altered the draft letter and draft reply to deal specifically with the alleged contact with Mr David Walker of KMS Limited, rather than with "any member" of KMS Ltd.' The need to make this amendment strongly suggests that Thatcher did at least have some contact with KMS members, even if it was not David Walker himself. By narrowing the scope of Thatcher's answer, they could deny Foulkes' allegation while also throwing him off the scent.

Meanwhile, another diplomat was coordinating a study of FCO contracts with KMS Ltd, which would be Whitehall's fourth such dossier on the firm. The results thus far were disturbing. There was no evidence of contact between Number 10 and Walker, 'but it is clear that the Prime Minister has in recent years met Mr Tim Landon, the former equerry to the Sultan of Muscat, who appears to be associated with KMS even if he may not be a member of the Board of Directors'. This connection, between Thatcher, Landon and KMS, was potentially explosive. The diplomat flagged up the link to Thatcher's trusted aide, Charles Powell, who wanted to check with her how to respond.[11] Was the White Sultan's role in KMS about to emerge?

Powell's intervention delayed a response and appeared to annoy the diplomat who had noticed the connection between KMS, Landon and Thatcher. Less than ten days later, the same official reminded their superiors: 'While we were not aware of any contact with Major Walker, the Prime Minister did know another associate in KMS, Mr Tim Landon.' A colleague responded with a handwritten note at the top of the memo: 'Thanks. Used briskly with the PM yesterday. No new insights.'[12] The language suggests that either the diplomat's concerns about Landon were not taken seriously, that they were brushed off to prevent scrutiny, or that even Thatcher's most trusted personal aides were themselves struggling to get 'insights' from the Iron Lady about her real relationship with KMS. As it happened, this ultrasensitive connection between Thatcher, Landon and KMS would only emerge over 30 years later when files from this episode were declassified at the National Archives in Kew. If the records are accurate, and Tim Landon was a secret 'associate in KMS', then the ramifications are profound. Compared to Landon, Walker now looks like a supporting character.

Landon is significant not merely because of his power in Oman, but also because of his deep personal connections to Margaret Thatcher, which had become something of an issue earlier in the 1980s. Although Landon's role with KMS remained unknown, there was public concern about his relationship to Mark Thatcher, the prime minister's son. As part of the Sultan's infrastructure drive to pacify the rebels in Oman, he awarded a lucrative contract to a British company, Cementation International, to build a university. This firm was a subsidiary of Trafalgar House, a property, construction and engineering conglomerate whose consultants included Mark Thatcher. The issue emerged in the press in 1984 and was seen as a potential conflict of interest, with some saying Mrs Thatcher had made use of Britain's alliance with Oman to win her son a business deal.

According to her private secretary at the time, Robin Butler, Thatcher's conduct around the deal 'conveyed a whiff of corruption'. He later told Thatcher's biographer: 'She had wanted to see Mark right. She sought the deal for Mark. She excluded everyone from her talks with the Sultan. Mark was dealing with Brigadier Tim Landon who was the Sultan's go-between. She behaved in a most peculiar way. I suspected the worst.'[13] A *Private Eye* piece in 1984 also raised questions about 'Tim Landon's access to Number 10', causing concern among British diplomats dealing with Oman.[14] As well as the popular satirical magazine sniffing around the White Sultan, TV investigators from *World in Action* were also on his case. 'Landon is aware of this', a diplomat commented, 'and has been trying to pull every possible string to put the TV researchers off: he has been in touch several times with No.10, but there is of course little we can do to stifle the programme'.[15]

There appears to be no doubt that Landon did have access to Downing Street. In 1984, the White Sultan met Thatcher's private secretary, Robin Butler, to deliver two messages about British candidates for the Omani civil service and their next head of defence. A letter about this meeting shows that Landon was treated with the utmost respect by Downing Street and that his views on the matter were taken most seriously.[16] Given that we now know Landon was also an 'associate in KMS' who had access to the prime minister for something as delicate as helping to win her son a business deal, this lays bare for the first time the truly intimate connection between KMS and Downing Street. The extent of this relationship did not become clear at the time because of Parliament's focus on David Walker, who MPs assumed was the senior partner in the whole affair. Yet the company's most powerful associate was the White Sultan, a kingmaker with the keys to Muscat and the ear of Mrs Thatcher.

*　*　*

The turbulence in Westminster did little to disturb the KMS contract in Oman. Sultan Qaboos had approved the company's suggestion of a motto ('Swift and Deadly') for his Special Forces unit, which had grown to include three highly mobile fighting squadrons that could roam around Dhufar in Land Rovers, and a counterterrorism squadron ('the Cobras') ready to storm any hijacked aircraft or ship. And despite the furore over KMS links to Whitehall, the company was still able to take the Sultan's Special Forces on a training trip to the main SAS base in Hereford for counterterrorism drills – as well as going on maritime security exercises in Hong Kong with the British colonial police Special Duties Unit. The training was necessary because although the threat posed by Dhufari revolutionaries had significantly dissipated by the late 1980s, it had not completely disappeared. Oman's intelligence agency had to call on the KMS-run Sultan's Special Forces to stage Operation Thalib, a covert raid on a rebel arms cache which contained 'a veritable Aladdin's cave for sedition', replete with pamphlets, guns and rocket-propelled grenades. The unit's second in command, KMS officer Alastair MacKenzie, later recalled that his men found 'a complete set of equipment with which to start a revolution!' The existence of this mission, revealed in MacKenzie's 2019 memoir, *Pilgrim Days: From Vietnam to the SAS*, confirms that KMS did play an active role in repressing Dhufar's liberation movement, denying it the means to re-launch an armed struggle.

Throughout the late 1980s, the company ensured that this elite unit remained under the command of a British SAS veteran, most probably Colonel Keith Farnes, but KMS did hire an increasing number of white Commonwealth mercenaries to serve as officers, including an Australian Vietnam veteran, several Rhodesians and at least three New Zealanders such as MacKenzie. He claims that among his compatriots hired by KMS in Oman was Ron Mark, a half-Maori/half-Irish military engineer who spent several years running the Sultan's Special Forces' vehicle workshop and rose to the rank of major. Mark has since said: 'It was my job to ensure they had what they needed, when they needed it.' When I communicated with Mark via email in 2019, he confirmed that he was 'initially contracted by KMS' in Oman, but said that at some point, possibly around 1987, the Sultan's Palace Office started to employ him directly. Mark's time with KMS, however brief, is important because after leaving Oman in 1990 he returned to New Zealand where he was elected as an MP in 1996. At the time of writing, in 2019, Mark had risen to become defence minister in Jacinda Ardern's Labour-led coalition government, arousing controversy when he proudly wore his three Omani service medals at official functions in breach of protocol. This led New Zealand's largest newspaper to publish an old photo of Mark in his Omani uniform shaking hands with Sultan Qaboos. But beyond the medal fiasco, the minister has escaped broader criticism over his work for a company that was embroiled in the Iran–Contra scandal, the massacres of Tamil civilians and the propping up of a dictatorship. When I challenged the minister on this point, he said he was 'unaware of KMS's activities outside of Oman' at that time, and that although he was 'contracted for a short time to KMS he was always under the command of the Palace Office'. Apparently, the minister 'considers his work in Oman service for Oman, not KMS'.

* * *

His home was easily identifiable from the car parked on the driveway, bearing a personalised number plate with his name. A British flag flew from an enormous flag pole on the front lawn. With my colleague Lou Macnamara filming over my shoulder, I knocked tentatively on the front door, which had been left ajar. No one answered, and after over ten minutes of waiting for a response we were beginning to give up. My other colleague, Angus Frost, was already at the end of the driveway, almost at the road. Suddenly, we heard a noise from an upstairs window. Frost and Macnamara quickly panned their cameras up, as I called out a tentative 'Hello'. A well-built man in his eighties, balding and bespectacled but still recognisable as Baty from the old army photos we had seen of him, cracked open a rotting wooden window. 'Yes and you can bugger off!' He shouted down at me, before I had a chance to introduce myself. 'Are you Lieutenant Colonel Baty?' I countered, trying to sound calm.

'You can bugger off!' He repeated, irate.
'Are you Brian Baty sir?' I persisted.
'You can bugger off!'

'Can I ask you some questions about your time … '
'No!' He shouted over me.
' ... with KMS in Sri Lanka?'
'No!'
'These are very serious allegations sir.'
'Bugger off my land, get out!' he shrieked.
'I've written to you sir as a journalist.'
'Get out!'
'You're being linked to war crimes in Sri Lanka.'

'You're a crackpot, conspiracist!' Baty muttered as he began closing the window, moving away from view. 'Do you have any comment about these allegations sir?' I responded, realising that any chance of an interview had literally gone out of the window. 'Were you the most senior KMS man in Sri Lanka?' I asked rhetorically. 'Do you accept any personal responsibility for war crimes that happened in Sri Lanka, sir? Do you deny that KMS was involved in war crimes in Sri Lanka, under your watch?' By now Baty had stepped away from the windowsill and disappeared from sight. Baty clearly did not want to be interviewed, and had already ignored the earlier letters I had sent him asking for an interview and outlining all the allegations against him.

In fact, this little encounter in a sleepy English village had caught the attention of the SAS, who were clearly alarmed that a member of its old guard was being approached by a journalist – even if the story was about what they had done since leaving the SAS, not while they were in it. After my abortive visit to Baty's home, a serving senior non-commissioned officer from the regiment contacted Horsfall and asked if he knew me, and if he knew I was trying to talk to Baty. They then tried to remind Horsfall about 'loyalty' to the regiment and tried to find out what Horsfall had told me. Horsfall was unimpressed, and told the caller he should not really be reaching out like this. Horsfall said he had already given me an interview, in which he told the truth, and that he would talk to whoever he wished. This extraordinary call lends further credence to the sense that there were close connections between the SAS and KMS.

If Baty were to one day respond more substantively, he might argue that KMS was helping a legitimate government fight a terrorist organisation. I put this hypothetical point to Horsfall, who handed in his notice to Baty back in 1986. 'I left because I wasn't happy with the role that I was being asked to fulfil', Horsfall told me. 'There were a lot of undercurrents that were immoral. They were wrong. It doesn't matter if somebody legitimises a government or recognises a government, it doesn't make them right.' Horsfall explained:

> There's been a terrible genocide that's taken place in Sri Lanka over the past generation, the Indian government have tried to stop it, the Tamil Tigers have tried to stop it, and nobody has succeeded. And now the Tamils are a very tiny minority, and a frightened minority that still exist in the country under fear and I'm frightened that eventually there are going to be no Tamils in Sri Lanka at all.

Horsfall's moral stance was at odds with his colleagues, men like Baty who stayed in Sri Lanka for another two years even as the situation continued to deteriorate. Horsfall suggests how it was possible for other KMS men to take a different approach to him. 'Well I suppose if you're running a company, a company's job is to make money. If you're an individual, then you have to have the integrity to take your own moral stance.' These men would have weighed up how much they needed the money, and whether it was worth staying on in order to pay their bills. Tim Smith faced a similar idea as he wrestled with the morality of his work in Sri Lanka, but ultimately went back for a second tour, largely so he could claim a year of being non-domiciled in the UK and reap the benefit of not having to pay tax on his earnings. 'Tell that to some kid's mother as he is finally laid to rest', Smith said, bitterly disenchanted with his decision. 'Tell her he died for a hole in some Englishman's roof; he died because of the British government's vague idea of political influence in the region.'[17]

Although Smith was initially a 'reluctant mercenary', tricked by KMS into flying combat missions and pressurised into captaining gunships, in the end he had relished the challenge. He coached less experienced pilots how to unload troops while under fire, how to manoeuvre the helicopter so the door gunners could give covering fire to ground troops and how to fly the helicopter so the pilot could accurately fire the devastating pod gun rockets. Without Smith's dedication to the job, and that shown by his KMS colleagues, the Tamil liberation movement may well have survived the onslaught of the mid-1980s and established an independent state, before the cycle of violence span completely out of control.

After Smith finally left Sri Lanka in April 1987, he returned to the UK and flew helicopters across the North Sea, servicing offshore gas rigs. However, he was plagued by flashbacks from his time in Sri Lanka, which were only made worse after he had a motorbike accident in 1995. In his memoirs he gives a taste of the nightmares that plagued him over a decade later: 'Old men, like scarecrows, still hang grotesquely tangled in the branches of a tree. Its all still there, locked in my mind.' The post-traumatic stress disorder took an immense toll on him, leaving him 'soul-stirringly sad'. He knew the horror of what he had done. 'So many Tigers tried to save their small piece of humanity. I nailed them to the ground.'[18]

*   *   *

After its exposure in front of the glare of the United States Congress, KMS was no longer able to operate at will around the world. However, its operations in Sri Lanka over the three and a half years since 1984 had done enough to prop up the government. By mid-1987, instead of the Tamil movement celebrating a year of liberation, it was on the point of military defeat, and had to rely on Indian intervention to provide it with enough breathing space to survive to fight another day. Delhi averted a bloodbath, dropping humanitarian supplies by air onto Jaffna, before signing a formal accord with President Jayewardene on 29 July 1987 that temporarily ceded the north and east of Sri Lanka to an Indian Peace Keeping Force (IPKF).

Many Sinhalese in the south of Sri Lanka were then horrified at what they saw as an Indian invasion on behalf of the Tamils, confirming their prejudices that Tamils were really just Indian settlers. This resulted in a chauvinist backlash against President Jayewardene, led by none other than a revitalised JVP, which had originally risen up in 1971. While some of the JVP's more enlightened left-wing cadres (such as the *Hiru* group) urged the JVP to form an alliance with the Tamil revolutionaries and create two Marxist states on the island, something PLOTE had long advocated, the JVP high command could not overcome its racism towards the Tamils. They launched a guerrilla war to try to topple the Sri Lankan government, which they believed had needlessly given ground to the Tamils.

Faced with this new insurgency, the exclusively Sinhalese STF proved just as willing to kill its own kith and kin in the south as it had slaughtered Tamils in the east. David Gladstone, who was Britain's ambassador in Sri Lanka throughout the second JVP uprising, later told me that Sinhalese villagers had complained to him bitterly about the STF, 'who they accused of carting people off, and disappearing them into big camps, which were not really acknowledged as camps, and from which one felt as if a lot of these people never really re-emerged. But it was all cloak and dagger stuff – it was not acknowledged. Everyone knew it was happening but nobody could pin it down.' Gladstone added that the STF and other units 'just went on killing and killing, until at a conservative estimate there were 60 odd thousand young Sinhalese men who were killed by the security forces' by the end of the 1980s.

Meanwhile in the north and east, the Indian peacekeepers turned out to be just as ruthless as the Sri Lankan forces that they had replaced. The IPKF managed to co-opt some of the armed Tamil factions, who increasingly became controlled by Indian intelligence, to wage a war against the more autonomous LTTE, which refused to lay down its arms or accept Delhi's interference in its affairs. In what became an increasingly fratricidal conflict, there was one side that stayed remarkably well hidden. Although most assumed KMS had shut up shop after the Contra affair, in fact it had never left Sri Lanka. And despite the increased Indian influence over the island, a power who had long opposed the presence of British mercenaries, the company found unlikely friends. KMS pilots flew SLAF helicopters in support of IPKF operations in the Jaffna peninsula until 27 November 1987, when the company's last pilot was withdrawn. Gladstone noted ironically in a telegram at the time: 'In the circumstances, it is not surprising that we have heard no recent criticism from Indian sources of KMS activities.'[19] The Indian government had gone from hating British mercenaries, to using them for aerial attacks on LTTE strongholds, over a four-month period in which Indian sources themselves estimate 1,000 Tamil civilians were killed by the IPKF. These airborne operations also included a concerted effort to wipe out the militant LTTE movement and eliminate its leader, Prabhakaran.

Simultaneously, KMS was also keeping busy in the south, where the JVP had attacked Sri Lanka's Parliament with a grenade, prompting the company to advise on improving security around the seat of power. KMS flew out an expert in this field, Willie Wilson, who produced a study for the country's defence secretary.[20] The ongoing unrest meant Brian Baty was scheduled to stay in Sri Lanka until April 1988, while he put in place an advanced command course at the STF training camp to prepare for the day he left. The company also maintained a former British army intelligence corps warrant officer 'within' the crucial Joint Operations Command, who was assisting in the 'collation of military intelligence' and must have played some role in suppressing the JVP.[21] A fixed-wing flying instructor and a helicopter guru were scheduled to remain teaching the air force 'indefinitely' at Trincomalee and Colombo respectively, as would five KMS trainers at the STF academy. Despite this formidable presence, the ambassador, Gladstone, felt it was a 'considerably reduced commitment' by KMS, from an all-time high of 42 to just seven after April 1988. And when Colonel Johnson called on him to deliver this good news, the mercenary 'intimated that the reduction had been brought about as much as anything by the [Sri Lankan] Ministry of Defence's lack of money'. It is unclear from the available archive when KMS finally left Sri Lanka altogether.

Meanwhile, Walker was busy rebranding the company. He incorporated a new entity, Saladin Holdings Ltd, that was registered to the same address on Abingdon Road. Walker was the largest shareholder of Saladin Holdings, which almost immediately took over Saladin Security, the entity that had existed since 1978, and had operated in parallel to KMS. It shared many of the same directors, men like Walker, Johnson, Nightingale and Wingate Gray, and operated out of the same office at 7 Abingdon Road. While KMS Ltd appears to have never been officially registered as a company in the UK (instead there are traces of it in Jersey and the Cayman Islands), Saladin Security seems to have been the UK-registered sister company. Saladin's website would later proudly describe KMS as its 'predecessor', claiming that together they had 'provided security services since 1975'.[22]

The purpose of setting up Saladin Holdings is unclear, but it did put distance between Walker and KMS, whose brand had been tainted by exposure in Congress, Parliament and the press. A Director's Report for Saladin Security explains that in May 1987 'the running of the company was taken over by individuals of commercial and military backgrounds'.[23] This description is bizarre, because it was already run by distinguished veterans and businessmen for the previous decade since its inception.

The report went on to say: 'This combination will expand the company's activities into corporate defence, offering services to financial institutions and international industrial and commercial organisations requiring the full spectrum of security advice.' However, this 'full spectrum' was in reality a pale imitation of what KMS had done. Henceforth, Saladin would concern itself with services ranging from 'information security, product contamination, foreign project and investment evaluation, to personal threat and kidnap'. Walker was aware that computers were becoming increasingly important, and Saladin would be an early proponent of cyber security. The directors tried to assure staff they would not be sacked, saying that they 'intend to consolidate staff to be compatible with the new corporate policy. The Directors anticipate that it will take at least two years for this new policy to be truly reflected in the company's turnover and profitability.' By the end of its first year of trading, Saladin Holdings recorded a modest post-tax profit of £34,537, which would now be worth almost £100,000.[24]

By trawling through Saladin's accounts, which I had to pay Companies House to digitise for me because they were so old, it is possible to identify another political connection that KMS might have benefited from. In November 1993, Archibald Gavin Hamilton joined David Walker as a director of Saladin Holdings Limited. Hamilton, a former lieutenant in the Coldstream Guards, was at that point the Conservative MP for Epsom, a Surrey town six miles from Esher, where David Walker had been a Conservative councillor. Hamilton had just stepped down as a defence minister earlier in 1993, a position he had held since 1986. In fact Hamilton had even answered a parliamentary question about KMS in March 1987, as pressure on the company was mounting over its work in Nicaragua. Labour MP George Foulkes had asked 'the Secretary of State for Defence if his Department at any time over the past five years has had contracts with KMS Ltd' – to which Hamilton answered, 'I know of none'.[25] His answer was truthful, but did not elaborate at all on the very real links between the Ministry of Defence and KMS, such as the frequent meetings Britain's defence attaché in Colombo had with Brian Baty.

Hamilton's appointment as a Saladin director is a classic example of the revolving door between public office and the private security industry. A senior politician who deflected scrutiny of a mercenary company had, on leaving office, taken up a role with its successor company. To add to the irony, Hamilton also took up a position on Parliament's Committee on Standards in Public Life, which was meant to scrutinise conflicts of interest. When I emailed him in 2018 to ask for an interview about KMS, he claimed: 'This is the first time I have ever heard of KMS Ltd. I am, therefore not qualified to say what KMS has or has not done in Sri Lanka and must decline your request to be interviewed on the matter.' After I pointed out that he had answered a parliamentary question about KMS in 1987, he said: 'I commend you on the quality of your research. However, I have no recollection of any reference to KMS while I was on the board of Saladin. Clearly the man you must interview is David Walker. If he will not talk to you, then neither will I.' For his part, Walker has simply ignored my emails asking for an interview.

Hamilton remained on the four-man board of Saladin until the end of November 1997. By then he was chair of the 1922 Committee, a powerful group of backbench Conservative MPs. Perhaps Hamilton left Saladin because its prospects were beginning to dry up, especially its most lucrative job of all. Saladin's financial statement in 1997 lamented: 'The group has seen a further decline in trading during the year with the Sultan's Special Forces in Oman', in a rare official reference to its work in the secretive Gulf monarchy.[26] However, Walker was not giving up, revealing that 'the directors and group's consultants have been attempting to open new markets, particularly in Africa, but no benefits have yet accrued'.

Walker's patience would eventually reap rewards. While Hamilton was appointed to the House of Lords in 2005, Walker's

African adventures took a little longer to pay off. He set up a series of subsidiaries across Africa, including one named Somalia FishGuard Ltd. In July 2013, Somalia's fishing ministry entered into an agreement with FishGuard, making it the 'sole agent responsible for the sale of fishing licenses' in Somalia. Crucially, in return for these efforts, FishGuard was granted 'a 49 per cent share of the revenues generated'. In other words, David Walker's company now owned half of Somalia's fish. For a nation sadly synonymous with piracy (many of whose practitioners were originally fisherman put out of business by European trawlers and nuclear waste dumping), David Walker had become the pirate in chief. FishGuard was supposed to monitor Somalia's fishing grounds, acting effectively like a privatised coastguard. Shortly before signing the contract, Walker's old friend, now Lord Hamilton, stood up in the House of Lords and asked a minister if she agreed that 'one reason why piracy has been reduced so radically is the use of private security companies on merchant shipping going through pirate-infested areas?'

Exactly as Walker had pioneered the use of modern mercenaries on land and in the air throughout the 1980s, he was now putting mercenaries out to sea, where they could patrol the oceans for profit. However, it has not been smooth sailing for Walker in Somalia. A year after winning the contract, Somalia's fishery minister, Mohamed Olow Barrow, wrote a letter criticising the terms of the deal and demanded that they be renegotiated. The UN Security Council reported in October 2015 that the deal had by then been suspended, and expressed concerns that the sale of fishing licences had the 'potential to fuel corruption and even conflict', especially as there was no agreement in place to share the revenues among different parts of the country.[27] Aside from the risk of corruption, the UN Monitoring Group noted wider concerns about privatised coastguards, warning: 'Somalia's past practice of entrusting private companies with the dual role of selling fishing licences and managing maritime security posed a threat to peace and stability. Such companies have privileged foreign clients, even to the point of providing foreign fishing vessels with armed guards, again leading to potential conflict with local fishermen and coastal communities.'

Notwithstanding this setback, Walker also won business in Somaliland, the self-governing territory in the north of the country. There, Saladin won a contract worth a reported US$6 million to train the country's new Oil Protection Unit (OPU), which was designed to reassure foreign oil companies that their assets would be safe. Similar to the STF in Sri Lanka, Somaliland's OPU is a police-military hybrid. The OPU would grow to 580 members across six mobile units, managed by Somaliland's interior ministry in liaison with risk-management firms, and hired by the oil companies. Its area of operations would include territories that were disputed with neighbouring Puntland, which led one observer to say the OPU's deployment would be 'a red rag to a bull'.[28]

The Horn of Africa is not the only place where Saladin has courted controversy. Like many private security companies, the conflict in Afghanistan has been an irresistible opportunity, and a country that David Walker is no stranger to. While KMS appears to have schemed in the 1980s to topple the Afghan government, from the 2000s Saladin supported efforts to stabilise the security situation in Afghanistan by offering bodyguards to foreign dignitaries – a service redolent of the early days of KMS. In 2002, not long after the fall of the Taliban, 'Saladin Afghanistan' registered with the Kabul government. By 2007, Saladin was reportedly providing heavily armed bodyguards for Canadian diplomats in Kabul, and even the country's then prime minster, Stephen Harper, when he visited.[29] The contract caused controversy in Canada's parliament, where politicians raised concerns about Saladin's previous life as KMS.

However, a spokesperson for Saladin Afghanistan insisted that the company's operations were above board, telling a reporter at the Canadian *Globe and Mail* newspaper: 'We run our Afghan operations in a professional discreet manner.' Saladin put the controversy down to 'jealousy and resentment' caused by the firm's success, and claimed without any hint of irony that 'having no skeletons in the cupboard we are in a very comfortable position'. Admittedly, by that point it was the newer private security firms, like the American company Blackwater, that were causing the most controversy. Blackwater operatives had killed 17 Iraqi civilians in September 2007 while guarding a diplomatic convoy. These fiascos elevated Walker to something akin to an older brother in the private security world: wise enough to keep his nose clean, and able to look on as young mavericks repeated mistakes he had made decades ago. Indeed, Walker is so confident that he keeps coming back to Afghanistan. In 2018, Saladin bid for an EU security contract in Kabul worth 100 million euros, although the deal was eventually awarded to rival firms.[30]

Saladin has also started to struggle in Saudi Arabia, where *Intelligence Online* claims that it 'is one of the most established security operators'. Although the company has, together with KMS, a long history of guarding Saudi royals, its fortunes are said to have tumbled under Crown Prince Mohammed bin Salman. Saladin has 'lost several of its long-term clients' in Saudi Arabia, including members of the royal family, and the company's 'books are beginning to sink even further into the red as it awaits payment for its latest services to these princes poor at paying their debts'.[31]

Walker is now in his mid-seventies, and it appears that a younger relative is now playing a prominent role in Saladin's affairs, possibly his son. While Saladin still uses the west London address that once recruited KMS pilots for the killing fields of Sri Lanka, the elder mercenary has long listed his address as an affluent expat suburb of Nairobi called Karen – one of the most exclusive parts of the East African capital. Whether or not he really lives at this address is unclear. When a friend of mine offered to check it out, he was surprised to find the taxi driver delivered him to a hotel-cum-wedding venue.

What is certain is that David Walker has activities in Nairobi, and provides armed guards for an anti-poaching charity in Kenya called the Lewa Wildlife Conservancy. This is effectively a rhino charity, which manages a vast wildlife conservancy, sparking resentment from some local Kenyans at the continued control of land by white expats. However, the legacy of KMS goes deeper than just David Walker's subsequent career path. By the time KMS rebranded as Saladin in the late 1980s, the company had blazed a trail of creative destruction and carved out a new industry. Although the company has not engaged in large-scale atrocities since the 1980s, that role has been taken on by other mercenary companies, such as Sandline and Executive Outcomes in the 1990s, Blackwater and G4S in the 2000s and the Spear Operations Group now operating in the war with Yemen. Both Westminster and Washington have been loath to introduce state regulation, and 'private military security contractors' have flourished in a zone of 'self-regulation'. Now, London and Hereford are global hubs of the mercenary industry. Had Parliament taken a different approach in 1987, and launched an inquiry into KMS's operations in Nicaragua and Sri Lanka, or if the police had probed David Walker and Brian Baty, then the industry may have looked far less attractive to other SAS veterans. Instead, Britain's premier special forces regiment has become a conveyor belt for more lucrative careers in the private sector, devoid of any official oversight.

Of the company's most senior staff, Colonel Johnson and Brigadier Wingate Gray have long since passed away, but there is still a slim chance that surviving members such as Walker and Baty could one day face repercussions for their work in Sri Lanka. The Metropolitan Police's War Crimes Unit could decide to launch an investigation, given that alarming evidence of their links to mass atrocities has only recently been declassified. However, the mood in British politics is currently moving away from any efforts to prosecute veterans for historic crimes. The Ministry of Defence has signalled it would support a statute of limitations on any offences involving armed forces personnel that occurred over a decade ago, amid fierce criticism of Northern Ireland and Iraq war veterans facing legacy litigation. In this climate, the prospect of veterans being prosecuted in the UK for offences they may have committed after moving into the private sector seems remote. The countries that have been most effective at prosecuting British mercenaries are those in the Global South who have been on the receiving end of private security contractors. In 2011, an Iraqi court sentenced a G4S employee, Danny Fitzsimons, to 20 years' imprisonment for murdering two of his colleagues, in what was the first conviction of a Westerner in Iraq since the 2003 invasion.

Yet isolated action against a few individuals is unlikely to signal a ban on mercenaries as a whole, so long as the British army continues to recruit large numbers of foreign fighters into its own ranks. This pervasive practice makes any meaningful definition of a mercenary legally fraught for Whitehall. There is the specific issue of the British army's Gurkha brigade, which consists of 3,000 Nepali soldiers. Each year, hundreds of Nepal's fittest young men trek to the Himalayan city of Pokhara to undergo a gruelling selection course, in the hope of earning a better life in the British military than they might otherwise have at home. Many of the Gurkhas end up in Brunei, where their garrisons prop up another autocratic pro-British Sultan. In 2019, a record 400 new Gurkhas were allowed to join the UK army, as part of a massive recruitment drive to make up for a fall in the number of British youths enlisting. The disastrous wars in Iraq and Afghanistan have made life harder for army recruiters, and the Ministry of Defence has increasingly looked to hire foreign fighters to fill the gap. Nepal is merely one source of willing mercenaries. In November 2018, the Ministry of Defence announced that an extra 1,350 Commonwealth citizens could join the military each year, regardless of whether they lived in the UK. This increasing focus on recruiting foreign fighters raises the question of how many British soldiers are actually British, or put differently, how many mercenaries work in the British army?

The Ministry of Defence does not just rely on foreign fighters, it also exports its own personnel as mercenaries to other countries. A parliamentary report published in 2016 found that there were 195 UK service personnel on loan to the Omani military, showing that the long-standing practice has not declined. All of them would fall into Lord Diplock's definition of mercenaries, as would the Gurkhas and other Commonwealth recruits. Ultimately, it seems as though mercenaries as an institution are so fundamentally British it is almost impossible to outlaw them. While the KMS men may have escaped any sanction over their operations, Margaret Thatcher's own son, Mark, pleaded guilty in 2005 to breaching South African anti-mercenary laws. He was accused of hiring a helicopter that was to be used in the so-called 'Wonga Coup' in Equatorial Guinea. Another pillar of the British establishment, Winston Churchill's grandson Sir Nicholas Soames MP, is a non-executive director of private security giant GardaWorld and the former chairman of Aegis, a mercenary firm that received over a billion dollars from the US government for its work in Iraq and Afghanistan. As such, the two most prominent British prime ministers of the twentieth century, Thatcher and Churchill, both had close family members who participated in the mercenary industry. Accordingly, it would take a truly determined occupant of 10 Downing Street in this century to finally put a stop to this very British business. Until then, national liberation struggles, progressive causes and concerned citizens will have to challenge mercenaries as best they can, wherever in the world private armies appear.

There was a missed opportunity in 2002, when the last Labour government produced a Green Paper on regulating the private security industry, which considered various options ranging from an all-out ban on mercenaries to a licensing arrangement. That move towards greater control of private armies was eclipsed by the invasion of Iraq, where the industry boomed, and by 2009 then Foreign Secretary David Miliband had done a volte-face, announcing that mercenaries should merely sign up to a voluntary code of conduct. It was another example of the inaction that Whitehall adopted in the face of the Diplock report, which would allow the existence of this ambiguous pool of ex-servicemen to continue, ready to take on jobs where the Foreign Office did not want to appear directly involved. Ultimately it would seem that for as long as British governments wish to intervene militarily in the affairs of other countries, mercenaries will remain an important tool in their arsenal, to be used in the most sensitive circumstances where Parliament, the press and the public would not stomach official British involvement. As such, any legislation that reigns in private military companies would also have the effect of constraining British foreign policy-makers from dabbling in secret wars. And perhaps that is why mercenaries are unlikely to be outlawed any time soon.

Epilogue

An unusual crowd gathered at Milltown Cemetery, where the sun shone brightly beneath a vast grey cloud bank. From afar, you could hear ritual Tamil singing echoing from a small sound system hastily erected in one corner of the graveyard. If you moved closer, you would have seen around a dozen Tamil men and women lining the graves of IRA members who died during the Troubles. Interspersed among the crowd was a similar number of local residents, mostly Sinn Féin members who had helped to organise the event. This unusual group was not there, however, to mark the deaths of the IRA volunteers lying in the earth around them. Instead, they were commemorating eight years since the end of Sri Lanka's war.

For many in the Tamil community, the war's conclusion in 2009 was a devastating demolition of their freedom struggle. Corralled into a shrinking corner of their island, the Sri Lankan army and air force relentlessly shelled civilians and combatants alike – and this time India did not intercede on their behalf as it had done in 1987. Supposed 'no fire zones', where Tamils were told they would be safe, were in fact killing fields – like the napalm incident at Thikkam Temple, but this time on a much larger scale. The commander of Sri Lanka's air force in the final stages of the fighting in 2009 was Roshan Goonetileke, who had learned to fly in 1986 with Tim Smith. He had risen to become an air chief marshall, and after the war would be chief of the defence staff, and then an adviser to Sri Lanka's president. There are many other young pilots schooled by Tim Smith who later went on to lead Sri Lanka's air force during appalling periods of aerial bombardment against Tamil civilians.

As many as 146,000 Tamils went missing in the final months of the fighting in 2009 – a figure compiled by a leading Tamil bishop, building on the tradition of the clergy who for so long had counted the dead. The Sri Lankan government vigorously disputes this statistic and initially claimed that not a single civilian was killed by Sri Lankan troops at the end of the war. Inconveniently for Colombo, however, a UN investigation has said as many as 40,000 Tamil civilians may have died in the final months of shelling. An ad hoc panel of international law experts even published a report characterising the conflict as 'genocide', a judgement that Horsfall would agree with.

The legacy of KMS in Sri Lanka should not just be measured by how many of its protégés rose to senior military positions by the end of the war in 2009. From my research, it seems likely that had KMS not supported the Sri Lankan government from 1984 to 1988, then the Tamil liberation movement may have achieved its military objectives as early as 1986, thereby averting decades of further bloodshed. While the Tamil movement did make serious errors, most noticeably the fratricidal rivalry between its different armed factions, this alone would not have prevented it from securing a swift victory. By mid-1985, the Tamils had fought the Sri Lankan government to the negotiating table, and achieved temporary unity among their revolutionary leadership. The Tamils were militarily superior to the Sri Lankan ground forces, both the army and police, and in the words of one British diplomat looked set to turn them into 'mincemeat'.[1]

However, KMS shifted the war in two key ways: by allowing the introduction of STF commandos in the east, who decisively turned the Muslim community against the other Tamil-speaking groups in the region; and by giving the Sri Lankan forces the ability to project power from the skies, raining death on Tamil fighters and civilians alike. Had it not been for these interventions, the predominately conscript army in the north would likely have capitulated (around 5,000 had already deserted), and the Tamil–Muslim alliance in the east could have resisted the arrival of the police commandos, as it was doing until April 1985. Perhaps this view places too much emphasis on the role played by KMS, but many senior Sri Lankan military veterans from that time (including the country's former head of intelligence Merril Gunaratne) have also acknowledged the company's pivotal role.

In inserting itself into the Sri Lankan conflict, KMS permanently altered the path of history. After the company left, the war became a direct contest between the Sri Lankan government and the LTTE in the 1990s, with the JVP, PLOTE and other militant groups eliminated and Delhi chastened from dabbling in the island's animosity. With India now out of the picture, the UK government opted to play a more direct role in subverting the Tamil struggle, without having to worry about blowback from Delhi. By the mid-1990s, Britain's defence attaché in Colombo was a veteran of Oman and Northern Ireland, and claimed to be a disciple of Frank Kitson, one of the UK's seminal counter-insurgency experts. In 1997, the British army helped Sri Lanka set up a military academy on the island, to train its mid-ranking army officers. Photographs from the time show that a serving British army colonel was one of the college's most senior staff.

A high-ranking British military officer remained in post at Sri Lanka's army staff college until at least 2001, but by then the LTTE had overcome its military deficiencies, and was able to beat the Sri Lankan forces on the battlefield, even without air power. Sensing victory, the Tigers declared a unilateral ceasefire, and sought international recognition of the sprawling de facto state that was now under its control. Unfortunately, rather than finally accepting the Tamil right to self-determination, the UK government responded in February 2001 by banning the LTTE as a terrorist organisation, significantly undermining the Tamils' chances of being treated as legitimate actors (a move that a handful of British MPs, including Jeremy Corbyn, dared to vote against). However, the Sri Lankan government had to take a more pragmatic approach, and accept the balance of forces on the ground. In 2002, they signed a ceasefire with the LTTE.

In practice, after over two decades of fighting the Tamil liberation struggle had now achieved its goal. There was peace at last, and Tamils had space to focus on some of the more progressive and revolutionary aspects of their struggle, from women's liberation to cooperative ownership and full employment. It was not a utopia, but women walked the streets safely at night, and Tamils had a chance, for the first time in centuries, to run their own affairs. Tragically, this state of peace would not last long. The Sri Lankan side used the truce as an opportunity to rearm, just as it had done in 1985 during the Thimpu peace talks. Once again, Britain was only too willing to help with this process. A retired British military officer conducted a 'defence review' for Sri Lanka, and the British government authorised the export of scores of weapons, for use on air, land and sea. Sri Lankan officers from all three services were welcomed on prestigious courses at British military and spy academies such as Sandhurst, Dartmouth and the Defence College of Intelligence. Meanwhile, on the international stage, Britain's Foreign Office pushed hard, along with the US, for the EU to ban the LTTE under anti-terror laws, further delegitimising the liberation movement.

By 2006, the ceasefire was breaking down badly. The STF played a particularly prominent role in ceasefire violations, massacring five Tamil students in Trincomalee, and taking part in the murder of Tamil aid workers from the French charity Action Against Hunger at Muttur, on the southern flank of the geostrategic harbour. Tellingly, the war restarted with a massive push to eliminate Tiger positions around Trincomalee. Elsewhere in the Eastern Province, the STF used tactics taught by KMS, particularly small reconnaissance teams, to drive out the LTTE in gruesome fashion. Several members of the STF were later alleged to have raped dead female fighters in this operation. Although KMS played no direct role in the last stages of the war, a senior

STF officer has claimed that a former member of the company, John Garner, continued to provide refresher training for the STF right up until 2009.

What is certain though is that the British government was actively involved at the end of the war. With hospitals being shelled in February 2009, the UK Foreign Office sent senior policemen from Northern Ireland to visit Sri Lanka as 'critical friends'. They ostensibly gave advice on 'community policing', although one policeman told me they also conducted a review of Sri Lanka's public order capabilities. By the time the war ended in May 2009, with tens of thousands of dead Tamils, the STF put their losses from the entire conflict at 462. Although many more were maimed and injured, the unit's losses were still a fraction of its adversaries. The STF continues to operate in Tamil areas, now without any fear of encountering armed resistance. One Tamil asylum seeker described to me how just after the war ended in 2009, the STF dragged him away from his family home one night and tortured him horrifically at their camp. He later escaped and claimed asylum in the UK in 2014, although it took another four years, and considerable evidence from me and my colleagues about the STF's atrocities, before an immigration judge finally overruled the Home Office and said he could stay in the UK.

By 2017, some of the founding fathers of the STF, like President Jayewardene's son Ravi, had begun to pass away, and the unit was in a nostalgic mood, keen to commemorate its 462 members who were killed and the 792 injured in the decades of conflict. It convened a 33-year anniversary ceremony at Katukurunda, which David Walker learned about in advance. He wrote a letter to the current head of the STF, Senior Deputy Inspector General Latiff, who was one of the first men trained by KMS back in 1984. In the letter, Walker described the first commandants of the STF as 'great men and good friends who played a major role in winning the war in Sri Lanka, and I would not want the occasion to pass by without paying my own tribute to them'. The letter, which was emblazoned with Saladin's logo, went on to say that 'my partner Colonel Jim Johnson and I were privileged to be able to make a contribution to the war effort'. Walker confirmed: 'we arrived in 1983 on the invitation of President Jayewardene and remained throughout his time in office'. Given that Jayewardene was president until January 1989, this suggests that KMS stayed in Sri Lanka for 18 months after the arrival of the Indian peacekeepers and throughout most of the second JVP uprising.

Walker said modestly that 'our main focus was in assisting and advising on the formation of the STF, although we were also involved in other areas'. Despite all the war crimes the STF committed, Walker wrote in 2017: 'It continues to be a matter of great pride that I frequently heard STF described as the most effective force in the country'. And in case there was any doubt about the continuity between Walker's many companies, he added helpfully that, 'At the time we operated as KMS Ltd – Saladin Security is its successor company'. To conclude this remarkable letter, Walker said that he often thought about the first STF commandants, 'and I admire them all … May God bless those fine men'.[2]

* * *

Walker is not the only one to have a rose-tinted view of the STF. Britain's Foreign Office also holds a benign opinion of the police commandos. In 2015, UK authorities drew up a 33-page report analysing the public order capabilities of Sri Lanka's police. The Foreign Office has refused to release the report under the Freedom of Information Act, but it is understood to be an 'in-depth study' of the STF, and contained over a dozen recommendations to enhance its capacity 'to manage public gatherings and protests in an effective manner whilst respecting fundamental human rights'. The cost of compiling the report was almost £15,000, which was paid for by Whitehall's secretive Conflict, Stability and Security Fund, as part of the UK's overseas aid budget.[3] Although the report is still classified, it seems as though it formed the basis for the Scottish Police College to then begin training the STF, under direction from the UK Foreign Office and with the approval of Nicola Sturgeon's Scottish government. An early sign of this training was in September 2016, when David Walker's pen pal, Latiff, visited Scotland and was given an award by the force's deputy chief constable as part of an 'educational tour'. A month before this visit to Scotland, Latiff had been made commandant (i.e. head) of the STF.

As stated earlier, Latiff was among the first batch of recruits to join the STF when it was created in 1984 and he was trained by KMS. By 1985, Latiff was a sub-inspector in the STF, and he participated in several operations against Tamil independence fighters in the country's Eastern Province up to and including 1987, when the STF was accused of widespread murder, torture and disappearances. In 2007, Latiff commanded STF operations in Sri Lanka's Eastern Province. A report by the International Truth and Justice Project, based on testimony from STF whistle-blowers, alleges that the STF was involved in the extra-judicial killing of Tamil prisoners in that area and period, on the orders of senior officers. After one execution, the bodies were burned with petrol while STF members ate a meal and drank *arack*, a local form of whisky.[4] It is unclear what, if any, vetting the Foreign Office did to assess Latiff's suitability for receiving UK assistance.

Since the classified study on the STF was drawn up in December 2015, Police Scotland have been training members of the unit almost constantly, and at considerable financial cost to the British public.[5] In addition to the STF commandant visiting Scotland, another senior police commando has made the trip – Assistant Superintendent Palinda Wijesundara, who went on to run basic training for hundreds of new recruits at the STF academy in Katukurunda. This officer takes a dim view of protest, describing post-war demonstrations as a 'nuisance' that had got 'out of hand' because people were taking to the streets 'for very small things'. However, he was concerned that the police's response, with water cannons and batons, was 'not good to the image of the police department', and claimed this was Colombo's motivation for receiving Scottish training.

Despite all this British support, the STF continues to be involved in human rights abuses. On 21 June 2017, the STF violently assaulted university medical students occupying the health ministry building in Colombo. Television channels broadcast footage showing hundreds of STF personnel wearing helmets severely beating students and stampeding them down the stairs. STF officers also baton-charged students staging a sit-down protest and used tear gas and water cannon to disperse the demonstration. An Inter-University Student Federation activist told the media that over 90 students were being treated at the national hospital, and some were in a critical condition. The following month, I visited the Scottish Police College at Tulliallan Castle and met a civilian manager in their International Development Unit who refused to watch the footage of the STF assaulting the medical students in Colombo.[6] When I asked him if he knew much about the history of the STF, he simply replied, 'I don't' – which again raises concerns that Police Scotland has not considered the STF's human rights record or history of failed British assistance since its inception, as it is required to do under UK government guidelines. If he had checked the archives, he might have spotted the warning from the British defence attaché who visited Jaffna in 1983, and accurately predicted that no amount of training could overcome the Sri Lankan security forces' inherent dislike for the Tamil population, which he felt 'cannot be corrected and will therefore continue to be an Achilles heel'.

As if to prove this point, days after my visit to the Scottish Police College a 17-year-old Tamil boy drowned after coming into contact with the STF. According to a complaint filed by the Asian Human Rights Commission (AHRC), Sathasivam Madisam was swimming in a lagoon with his brother and several friends when a large numbers of STF armed officers approached. The commandos started shooting into the air and the boys ran away as fast as they could. Sathasivam and his brother ran along the bank of the lagoon and both fell into the water and got into difficulties. Though several boys tried to rescue Sathasivam, the STF are said to have prevented them.[7] The pair were later taken to hospital – where Sathasivam was declared dead. On 1 September 2017, I met his family in their home in eastern Sri Lanka, and recorded an interview with his father, whose account corroborated the AHRC complaint. The killing sparked a vigorous protest and the victim's relatives maintain that he was extrajudicially killed by the STF officers.

That death was just one incident in a summer of discontent for the Tamil community. In August 2017, the head of Sri Lanka's police, Pujith Jayasundara, announced a crackdown on Tamils in the Northern Province, who were protesting another killing, this time by regular police (not by STF). Jayasundara promised combined STF/army patrols and searches 'until we arrest all responsible wrong doers, the bad characters'.[8] The round-ups were front-page news in the local media. When I visited Jaffna in August 2017, a local Tamil journalist said hundreds of STF members had gone house to house in a village called Thunnalai and arrested people who had 'nothing to do' with the earlier protests. He said the round-ups severely disrupted school A-Level exams. Upon hearing that Police Scotland claimed its training would improve the STF's human rights record, he said: 'It is ridiculous and it is makes me sad to hear that they make these claims.'

At the end of that summer of repression, Police Scotland instructors returned to Sri Lanka and gave STF instructors training in evidence-gathering and liaison skills, which could enhance the commandos' ability to gather intelligence on protesters. Dozens of Scottish officers were teaching in Sri Lanka the following spring, amid a series of riots against Sri Lanka's Muslim communities in which the police failed to control the situation and the government declared a state of emergency. One Muslim man was killed and hundreds of Muslim-owned properties, including mosques, were set alight. The rioters were from the country's Buddhist majority, and several monks were arrested for allegedly playing key roles in the violence. There is disturbing evidence that the rioters were aided and abetted by the STF. A Sri Lankan government minister, Abdul Haleem MP, highlighted an incident where 'people who were prepared to safeguard a mosque were chased away by the Special Task Force and then they gave an opportunity for the rioters to come and attack'. There are also allegations that in some cases the STF actively attacked Muslims. Journalists for Democracy in Sri Lanka (JDS) allege that the STF beat Muslim men and tried to frame them for the rioting. Abdul Saleel Mohamed Fazil, a 43-year-old local councillor, told JDS that the STF planted petrol bombs on him, bound his hands and feet and beat him with wooden poles. Mohamed Masood Faizal, 40, was also dragged from his house and beaten by the STF. 'Since the day of that incident, my kids wake up every night screaming', he told JDS. In response, Labour's shadow development minister, Kate Osamor MP, told me that it 'cannot be right if the British public's aid money, meant for poverty reduction, is instead supporting riot control training, without the checks and balances in place to prevent attacks on minorities and human rights abuses'.

These anti-Muslim riots fed into an emerging narrative since the Tigers' defeat in 2009, whereby hawkish elements in Colombo have sort to portray Islam as the new threat to the island. This has helped justify maintaining the bloated national security apparatus, and repositioning Sri Lanka more neatly within the global war on terror. To this end, there has allegedly been collusion between military intelligence and extreme Muslim and Buddhist groups, in what appears to be an attempt to fan a cycle of action and reaction. The 2018 riots were just one part of this puzzle, which grabbed the world's attention on Easter Sunday 2019 with the horrific bombings of hotels and churches across the island by a local Islamist cell, killing hundreds of innocent people. (The plotters hailed from near Batticaloa, an area once home to harmonious Muslim–Tamil relations until the STF drove a wedge between the communities in 1985.) Further riots against the Muslim community soon followed, in which the security forces were again seen colluding with Buddhist mobs to attack mosques. Despite this demonstrable lack of progress in Colombo's control of its uniformed personnel, Wijesundara claims that the training is scheduled to continue until 2020 and that Police Scotland has given the Sri Lankan police equipment for public order management, as well as providing specifications for other equipment that was available in China at a cheaper cost than the UK. Police Scotland insists it has not supplied any gear to the STF, but only gave them a list of British companies who made suitable pads, helmets, boots and shields.

Although the UK and Scottish governments have shown little concern about Police Scotland's work with Sri Lanka, there is one parliamentary committee at Westminster that has occasionally criticised the fund involved, which also finances UK police training in Bahrain and Oman. Bahraini activists have been particularly effective about raising concerns about the fund, and members of the committee pledged to hold an evidence session and question a cabinet minister about what was going on. Ironically, the former Saladin director Lord Hamilton also sits on the committee, and although he did not attend the hearing himself, the issue of Sri Lanka was not raised.

On occasion, the Tamil community has successfully challenged Britain's collusion with Sri Lanka's police. In September 2018, the head of Sri Lanka's police suddenly pulled out of a visit to the Scottish Police College, after news of his visited leaked and protests were planned. The famous Tamil musician M.I.A. blasted the botched visit as 'another example of [the] deep state to state military relationship between the UK and genocidal Sri Lankan state'. She told me the UK training bolstered the Sri Lankan forces against the Tamil struggle. 'In the wake of the Scottish independence referendum movement, it is terrible to see such relationships against Tamil self-determination being allowed to flourish on Scottish grounds', she noted, mindful of the Scottish independence movement's aspirations.

All the Tamils I have spoken to oppose use of British aid to support the STF. One widow, who had several of her family murdered by the unit in 1986, said: 'The British government has helped train the forces who are committing these atrocities. The British government needs to come down and see what they are doing. They need to investigate what is happening to us.' One refugee, who was tortured by the STF, said that it was not enough for Britain simply to grant asylum: victims of the STF should receive compensation as well. Perhaps it is time then that Britain's aid budget was used for reparations, rather than repeating the mistakes of the 1980s.

* * *

David Walker and Oliver North left Nicaragua severely scorned, but their surrogate army would ultimately succeed. The relentless war forced the Ortega-led Sandinista government to divert vast amounts of money towards military expenditure, simply in order to survive, leaving little money available for their socialist projects. When President Daniel Ortega held another election in 1990, after a five-year term, the country's voters knew a vote for him would only result in more US sabotage. Ortega was duly defeated at the ballot box, under threat of permanent war. The country's liberals and conservatives took power, but it was not the end of Ortega or the Sandinistas.

In 2006, Ortega won another presidential election in a remarkable comeback. At the time of writing, he remains president of Nicaragua, having won another two polls. However, many of his former allies on the left now question what he really represents. Some say he has softened his stance on US imperialism, and he has adopted a more socially conservative platform on issues such as abortion, where he now identifies more closely with the Catholic Church. In 2018, he was rocked by student protests against social security reforms, resulting in over a hundred fatalities. A smaller, but still significant number of police and Sandinistas are also said to have been killed by the protesters. However, while Ortega has moved closer to the US than in the Reagan-era, he is certainly not beloved by Washington. The National Endowment for Democracy, the US government's regime change vehicle, has poured millions of dollars into anti-Ortega groups in Nicaragua, leading some to claim that the 2018 protests are really a repeat of the CIA's destabilisation campaign of the 1980s. The outcome of these protests and future of Ortega and Nicaragua remains uncertain and hotly contested.

Unlike David Walker, who had to retreat from politics, Oliver North continues to be an active figure on the right of American politics. After fighting off the legal ramifications of the Contra scandal, he had a successful career as a military author and TV presenter, before becoming president of the National Rifle Association in 2018, meaning he now leads the most powerful gun lobby in the world.

Oman is in many ways the 'success story' for KMS, if measured against the company's own authoritarian terms. Sultan Qaboos, the man they sought to defend, remains on the throne today, a decade after the White Sultan himself succumbed to old age. Although almost nothing is known about Saladin's current activities in Oman, it almost certainly continues to train the Sultan's Special Forces in some capacity. While many regime apologists seek to portray Qaboos as a benign dictator, who was less cruel and more charitable than his father, such a conclusion not only sets the bar incredibly low, it also does an immense disservice to the Omani subjects who have had to endure almost half a century living under Qaboos' decrees. It is worth pointing out that his reign has been far longer than Colonel Gaddafi's in Libya or Saddam Hussein's in Iraq, and he is unquestionably the longest-serving autocrat in the Middle East.

Khalfan al-Badwawi, an Omani exile who has helped me immeasurably with research for this book, has told me what

Qaboos' governance means in practice: 'There are Omanis alive today who lived through this cruelty [in the 1960s and 1970s] and don't want to rise up again. It became normal to live with terror. So the regime has retained control using fear. Omani's cannot even say the Sultan's name and have to call him *abunah* – "our father". It is legally forbidden to say anything about the Dhufar Revolution.' When I asked him more about the infra-structure Qaboos built with Oman's oil wealth, al-Badwawi was scathing:

> The new Sultan built schools, but pupils are taught not to question anything. They designed new human beings who can't think, who have no intellectual curiosity – education was available but it was empty. We lived just to be loyal. We can't think for ourselves. We don't have political or civil rights. It hurts me to see Omanis who still live like this, without knowing anything about our past, or our future, we just live day by day, not allowed to speak out. And it makes me angry when I think that the British government designed and imple-mented this system for Qaboos.'

A sign that all is not well in Oman came in 2011, when the country proved it was not immune from the Arab Spring sweeping the region. Al-Badwawi joined the sit ins from day one, and felt the Sultan's carefully constructed 'wall of fear' suddenly collapse as protesters took over city centres. However, this Paris Commune-style situation did not last long and the army moved in, arresting al-Badwawi together with 800 others in his home city alone. He believes half a dozen protesters were killed in the clearance operation. Sticking to his role as the nation's benev-olent father, Qaboos offered amnesty to almost all the political prisoners and placated people with promises of employment. Sensing an opening, thousands of oil field workers, mostly Shell staff, soon went on strike and demanded rights to holidays. In response, hundreds were sacked and many arrested.

But something inside al-Badwawi had changed. He was no longer afraid of the regime, and began to challenge it more openly. He was incensed when Qaboos came to London amid the oil field strikes to celebrate Queen Elizabeth's ninetieth birthday, bringing an entourage of hundreds of horses. It was one of the largest horse transportations by air in history, costing millions – just a year after his people went out on the streets and died because they were asking for jobs. Al-Badwawi felt angry and humiliated, and organised a rare protest in the centre of Muscat asking who was more important – horses or humans? Thousands of masked riot police were sent to intimidate al-Badwawi's flash mobs, beating their shields until the protesters took the hint and went home. He was marked out for special treatment, and subjected to weeks of psychological torture in solitary confinement. He was tasered in the ribs with an electric shock baton, covered in a black bag from head to toe, shackled in leg irons and doused in pepper spray. Once in custody, he tried to kill himself, before deciding it was his destiny to survive.

Eventually he was taken to a court where the prosecutor and judge sat on the same table together, and faced charges of 'insulting the Sultan'. In the end he had 14 hearings before being released on bail, banned from working and barred from travelling abroad. While most of his comrades succumbed to the Sultan, al-Badwawi was determined to continue. The pressure on him was immense, but the situation was too stark to stop protesting. He heard Prince Charles was coming to Oman to promote trade deals, including the sale of Typhoon fighter jets worth billions and a BP gas deal where the British firm would take 60 per cent of the revenues. Al-Badwawi began to mobilise campaigners on social media, and before long the Internal Security Service – Oman's MI5 – kidnapped him two days before Prince Charles arrived. Masked men ambushed his car while driving, bagged him up again and put him back in solitary confinement. Al-Badwawi says his kidnappers used special forces-style tactics to seize him, and strongly suspects they were trained by Britain, possibly even KMS or Saladin.

He was released after Prince Charles left Oman, but by now he had no money and no work, and was constantly being followed and harassed. Eventually, he fled the country and sought asylum in the UK. Like the Tamil refugees who came to the former metropole, al-Badwawi was in for a rough ride in Britain. He spent two years in limbo, bearing the brunt of Theresa May's hostile environment, until an independent judge accepted he was 'at real risk of return to Oman because of his political opinion'. He still suffers from flashbacks and nightmares about his abuse in Oman, abuse that he made clear to the Home Office as early as 2014. Despite his warnings, Britain's Foreign Office decided the appropriate response would be to deliver more training to Oman's security forces. Between March 2014 and May 2017, Northern Irish police officers had eleven meetings with Omani security forces to enhance their capability to manage protests, under a scheme designed and paid for by the same dubious aid fund that allows the Scottish police to work in Sri Lanka. The Oman training has seen visits to Northern Ireland with sessions on 'the importance of intelligence for successful public order'. Recipients of this wisdom have included Major General Ahmed bin Ali bin Salman Kashoob, who was at that time head of Oman's National Security Council. Kashoob had previously been a member of Oman's Internal Security Service (ISS) from 1998 until 2013, rising to become assistant to the unit's chief. In other words, he held a senior position in the ISS when it was keeping al-Badwawi in custody on charges of insulting the Sultan.

Other visitors to Belfast have included the commander of Oman's Special Task Force (a police paramilitary unit similar to the one in Sri Lanka), a lieutenant colonel in the Sultan's Special Force (the unit KMS and Saladin originally trained) and the deputy commander of Oman's Northern Frontier Regiment, which opened fire on protesters at the start of the Arab Spring. The scheme had the blessing of the Democratic Unionist Party, whose leader, Arlene Foster, had dinner with the head of Oman's police at a five-star cocktail bar. Documents I have obtained about the scheme also revealed that a Northern Irish man, Lieutenant Colonel Dave White, was a consultant to the secretary general of Oman's National Security Council, part of Muscat's ultra-powerful Royal Office, suggesting that the days of White Sultans are not over yet.

As well as training Oman's security officials, the British military continues to use Oman as a giant version of Salisbury Plain. In October 2018, over 5,000 British soldiers played a vast war game in Oman called Swift Sword 3. The Royal Navy has opened a new base in the Omani port of Duqm, costing UK taxpayers millions of pounds, although the precise cost is classified. The British army is also about to open a training base just inland from Duqm. Effectively, Oman is virtually still under British rule. What will happen to the country when Qaboos eventually dies remains to be seen. It is hard to imagine that Saladin and all the other British military assets in Oman would welcome a democratic transition, and allow people like al-Badwawi back to their homeland.

Notes

Prologue

1. NorthEast Secretariat On Human Rights, 'Piramanthanaru Massacre', April 2006.
2. FCO 37/3930, F218, Alasdair MacDermott to Cabinet Office, 20 November 1985.
3. FCO 37/3930, F214, FCO SAD to PUSD, 3 December 1985.

Introduction

1. Sir Roger Carr, BAE Systems Annual General Meeting, 9 May 2019.
2. Leigh Day, 'The Kenyan Claimants', www.leighday.co.uk/International/Further-insights/Detailed-case-studies/The-Mau-Mau-claims/The-Mau-Mau-claimants (accessed 6 June 2019).
3. Duff Hart-Davis, *The War That Never Was: The True Story of the Men Who Fought Britain's Most Secret Battle*, Arrow Books, London, 2012, p. 78.
4. KMS Ltd rebranded and restructured under the banner of Saladin Security at the end of the 1980s.

Chapter 1

1. Abdel Razzaq Takriti, *Monsoon Revolution: Republicans, Sultans and Empires in Oman, 1965–1976*, Oxford University Press, Oxford, 2016 p. 192.
2. Anthony G. Miller and Miranda Morris, *Plants of Dhofar, the Southern Region of Oman: Traditional, Economic and Medicinal Uses*, The Office of the Adviser for Conservation of The Environment, Diwan of Royal Court, Muscat, 1988.
3. Takriti, *Monsoon Revolution*, p. 1.

4. Ministry of Information and Tourism (Oman), *Oman in Colour*, Muscat, Ministry of Information and Tourism, 1974.

5. Takriti, *Monsoon Revolution*, pp. 140–2.

6. Tim Landon, 'Dhofar Isolation Rebel Activity', 18 March 1970, cited in Takriti, *Monsoon Revolution*, pp. 140–2.

7. Takriti, *Monsoon Revolution*, p. 143.

8. Takriti, *Monsoon Revolution*, p. 139.

9. Takriti, *Monsoon Revolution*, p. 173.

10. John Beasant, *Oman: The True-Life Drama and Intrigue of an Arab State*, Mainstream Publishing, Edinburgh, 2013, p. 191.

11. Takriti, *Monsoon Revolution*, p. 208.

12. Takriti, *Monsoon Revolution*, p. 271.

13. Ministry of Information and Culture (Oman), *Oman: A Nation Builds its Future*, Ministry of Information and Culture, Sultanate of Oman, 1976.

14. Beasant, *Oman*, pp. 188-9.

15. Michael Barber, 'Brigadier Tim Landon', *Guardian*, 28 August 2007, www.theguardian.com/news/2007/aug/28/guardianobituaries. military (accessed 2 September 2019).

16. FCO 8/2961, British Ambassador in Muscat to Dr David Owen MP, 3 July 1971.

17. Calvin H. Allen Jr and W. Lynn Rigsbee II, *Oman under Qaboos: From Coup to Constitution, 1970–1996*, Frank Cass, London, 2000, pp. 91–2.

18. FCO 8/2961, British Ambassador in Muscat to Dr David Owen MP, 3 July 1971.

19. FCO 8/5511, annual report for 1983 by the British defence attaché, Muscat, 30 July 1984.

20. FCO 8/2961, annual report for 1976 by the British defence attaché, Muscat, undated.

21. Alastair MacKenzie, *Special Force: The Untold Story of 22nd Special Air Service Regiment*, I.B.Tauris, London, 2011, p. 309. See also J. E. Peterson, *Oman's Insurgencies: The Sultanate's Struggle for Supremacy*, Saqi Books, London, 2007.

22. 'Colonel Jim Johnson', *Telegraph*, 13 August 2008.

23. Robin Horsfall, interview with Phil Miller, Prague, 6 November 2017.

24. Hart-Davis, *The War That Never Was*, p. 80.

25. Michael Asher, *The Regiment: The Real Story of the SAS*, Penguin Books, 2008, pp. 396–7.
26. FCO 99/2620, 'Nicaragua/Blowpipes: Arms for the Contras', P. R. Fearn to Private Secretary, 25 March 1987.
27. FCO 99/2620, 'Nicaragua: Letter from Mr George Foulkes MP to Prime Minister', addendum by P. R. Fearn, 16 March 1987.
28. LinkedIn, 'Jason Manning', www.linkedin.com/in/jason-manning-ba69b445/?originalSubdomain=uk (accessed 6 June 2019). 'Saladin Security Ltd – Security Team Member for The Sultan of Oman providing security to Brigadier Tim Landon Dates Employed 2002–2007'.
29. Brigadier Mike Wingate Gray, *Herald*, 11 November 1995.
30. FCO 93/895, F266, Crosland to British embassy, Beirut, 18 November 1976.
31. Peter Dickens, *SAS: Secret War in South East Asia*, Frontline Books, London, 2016.
32. Howells studied Chinese at SOAS. He died of a brain tumour in 1984. He was head of Chancery in Poland from 1972 to 1975. He was head of the FCO Security Department by September 1975.
33. FCO 93/2447, 'Armed Bodyguards for Beirut Embassy', I. S. Winchester, FCO Internal Security Department, 20 June 1980.
34. Duncan Campbell, 'The Pedigree Dogs of War', *Time Out*, London, no. 433, 21–7 July 1978.
35. Duncan Campbell, 'The Pedigree Dogs of War', *Time Out*, London, no. 433, 21–7 July 1978.
36. CAB OD(80).
37. Dick Grogan, 'Paratrooper Took Aim and Shot at Me, Says Witness', *Irish Times*, 15 December 2000.
38. Hansard, 10 February 1976.
39. *Foreign Enlistment Act 1870* (Vic).
40. Hansard, 10 February 1976.
41. FCO 65/1768, D. H. J. Hilary to Martin Reith, 25 February 1976.
42. FCO 65/1768, Martin Reith to Mr E. E. Orchard MBE (Research Department), 26 February 1976.
43. FCO 65/1768, letter from Lord Diplock to Prime Minister Jim Callaghan, 12 April 1976.
44. Diplock, although embracing modern technology, was not anticipating the advent of smart phones and mercenaries uploading

videos of their activities to Facebook, as has happened in Syria since.

45. FCO 65/1768, 'Mercenaries: The Recommendations of the Diplock Committee', FCO West African Department, 6 August 1976.

46. FCO 65/1768, 'Record of a meeting held at the FCO on Tuesday 10 August 1976 to discuss the Diplock Report on Mercenaries', 10 August 1976.

47. FCO 65/1768, M. E. Heath, FCO West Africa Department, 30 September 1976.

48. FCO 65/1768, H. M. S. Reid, FCO Central and Southern African Department, 5 October 1976.

49. FCO 65/1768, I. T. M. Lucas, FCO Middle East Department, 2 December 1976.

50. FCO 37/4354, F59, 'KMS – FCO Involvement', A. C. Smith, 26 August 1986.

51. FCO 65/1768, C. J. Howells, FCO Security Department, 3 December 1976.

52. FCO 65/1768, undated draft.

Chapter 2

1. CJ 4/1304, 'The Shotgun', Undated.

2. WO 373/172/559, recommendations for awards and honours.

3. CJ 4/1304, B. M. Webster, Northern Ireland Office, 6 May 1976.

4. CJ 4/1304, 'Shooting of Peter Joseph Cleary', T. J. Oyler, 16 April 1976.

5. *Birmingham Post*, 'SAS Men in Court', 7 May 1976, p. 1.

6. Asher, *The Regiment*, p. 436.

7. CJ 4/1306, 'The Problem – Wider Deployment of the SAS', P. N. Bell to Burns, November 12, 1976.

8. Derek Brown, 'Dublin Sends Back SAS Wanderers', *Guardian*, 9 March 1977, p. 1.

9. Landon went to Sandhurst with the young Sayid Qaboos from 1960 to 1962, in the 29th intake. David Walker was there from September 1961 to 1963.

10. Roger Cole and Richard Belfield, *SAS Operation Storm: SAS Under Siege, Nine Men Against Four Hundred*, Hodder Books, London, 2012.

11. FCO 7/3043, 'Embassy Security', John Shakespeare to Chris Howells, 13 May 1976. It states Walker made a 'flying visit to Buenos Aires last weekend'. The coup took place on 24 March 1976.

12. FCO 7/2760, 'Bomb Explosion; HM Embassy Buenos Aires', and FCO 7/3043, 'Bodyguards for HM Ambassador Buenos Aires'. The latter refers to 'Whitehead's letter of 28 May 1975 to Creamer about the employment of UK-based bodyguards for the protection of our Ambassador at Buenos Aires'.

13. Bank of England, 'Inflation Calculator', www.bankofengland.co.uk/monetary-policy/inflation/inflation-calculator (accessed 6 June 2019).

14. David Walker's address at Colonel Jim Johnson's funeral, 30 July 2008, cited in Hart-Davis, *The War That Never Was*, p. 348.

15. FCO 7/3043, Untitled, Chris Howells to John Shakespeare, 27 August 1976.

16. FCO 7/3043, 'Security', John Shakespeare to H. M. Carless, 30 September 1976.

17. FCO 7/3043, 'Bodyguards at Buenos Aires', Chris Howells, 8 June 1976.

18. FCO 7/3043, 'Security', A. J. Sindall to John Shakespeare, 16 December 1976.

19. 'British Ambassador Assassinated in Holland', *On This Day*, BBC, 22 March 1979.

20. Cabinet, Defence and Overseas Policy Committee OD(80) Meetings and Memoranda, 1–3 Vol 1 H.58 p. 216.

21. Cabinet, Defence and Overseas Policy Committee OD(80) Meetings and Memoranda, 1–3 Vol 1 H.58 p. 216.

22. FCO 31/2699, 'Uganda: Presidential Protection Unit', A. J. Longrigg in FCO East Africa Department to Mr Rosling and Mr J. R. Williams, 19 April 1979.

23. FCO 31/2699, 'Uganda: Presidential Protection Unit', handwritten note about KMS by Mr J. R. Williams, 19 April 1979.

24. FCO 99/370.

25. FCO 99/370.

26. FCO 99/370.

27. FCO 99/370.

28. FCO 99/370.

29. FCO 7/2759, 'Mr Charles Lockwood', Hugh Carless to D. R. Ashe, 11 September 1975. It stated that MI5 'deals with precautions against kidnapping'.

30. FCO 99/370.
31. FCO 99/370.
32. FCO 99/370.
33. FCO 99/372.
34. FCO 99/372.
35. FCO 99/583.
36. Foreign Office FOI refusal, 0659-18, 25 September 2018.
37. CAB OD(80).
38. The *London Gazette* shows Walker was a reserve in the Royal Engineers and was made a Major on 20 November 1980.

Chapter 3

1. FCO 37/785.
2. FCO 37/787.
3. FCO 37/790, 'The Ceylon Insurgency, 1971'.
4. Hansard, 22 April 1971.
5. FCO 37/805.
6. FCO 95/1108.
7. Hansard, 6 May 1971.
8. Five-day visit with the British High Commission staff from 21 to 28 April. Colonel Roger May of the Royal Hampshire Regiment was selected, owing to his 'experience of insurgency situations in a number of parts of the world'. Colonel May's army career spanned occupied Germany, Palestine, Cyprus, Aden, Bahrain and Northern Ireland. The FCO sent their full-time 'Overseas Police Advisor', Mr Michael Macoun, to accompany Colonel May. Mr Macoun had trained in the Metropolitan Police and joined the Colonial Police in Tanganyika, where he had interned Nazi sympathizers. Later, as 'Superintendent of Police in Dar Es Salaam he had to cope with civil and industrial unrest resulting from the end of the war and demobilization'. During Macoun's tenure as inspector general of police in Uganda, he 'faced independence campaigns, rebellion, mutiny, [and Mau Mau] incursions from unstable neighbouring states'.
9. FCO 37/786.
10. FCO 37/787.
11. FCO 37/787, 'Ceylon Armed Forces – State of Emergency'.

12. Ranatunge would become chief of Sri Lanka's Defence Staff, defence secretary and then high commissioner to the UK from 1993 to 1995.

13. FCO 37/786.

14. FCO 37/787.

15. Fred Halliday, 'The Ceylonese Insurrection', *New Left Review*, vol. 1, no. 69, September/October 1971, pp. 84-5.

16. FCO 37/807.

17. Hansard, 26 April 1971.

18. FCO 37/786.

19. FCO 37/789.

20. FCO 37/810.

21. FCO 37/790.

22. Satchi Ponnambalam, *Sri Lanka: The National Question and the Tamil Liberation Struggle*, Zed Books and Tamil Information Centre, London, 1983, p. 63. While the shortcomings of the Soulbury Commission are often painted as the blunders of an aloof colonial administrator, oblivious to the complexity of local feelings, there is some evidence that British planners had a more deliberate motive in making the Sinhalese reign supreme in Ceylon. Throughout the struggle for independence, the Tamils had played a disproportionate role, with their Jaffna Youth Congress inspired by anticolonial unrest in India. It was the Tamils who had invited Mahatma Gandhi to visit Ceylon in 1927. And although the Tamils were not 'alien invaders' from India, they did share a common language with millions of Tamils who lived just across the Palk Strait in the southern Indian state of Tamil Nadu. The Tamils therefore were a potential linguistic conveyor belt for South Asian solidarity, between the Indian subcontinent and the island of Ceylon. This prospect troubled British imperialists, as Ceylon had been a useful 'fortress colony' from where the Crown could suppress dissent in India. To take just one example, in 1946, 20,000 Indian sailors in the Royal Navy staged a revolt, seizing control of over 70 ships and bases. It was by far the most serious threat to British rule in India, and did more to hasten independence than any non-violent resistance ever could. To crush the revolt, Britain swiftly dispatched a fleet of destroyers from the safety of Trincomalee harbour, which steamed north and regained control.

This desire to isolate Ceylon from unrest in India is reflected in a top secret memorandum produced jointly by Britain's chief of the air staff, the chief of the imperial general staff and vice chief of the naval staff in June 1947, in the midst of independence negotiations. They warned that: 'There is always danger of India (especially Congress India) interfering in the Ceylonese internal politics and promoting discontent among the powerful Indian minority', referring to the Tamils in Ceylon. 'Such disorders, however provoked, would have a serious effect upon the working of our [military] service establishments', the top brass noted, alluding to the Royal Navy base in Trincomalee harbour and elsewhere. The memo goes on to state ominously that, 'Although the Ceylon Government should be responsible for internal security, in the event of the situation becoming beyond her capacity to control and our defence interests being threatened, we should reserve the right to introduce forces, and to take action as necessary to protect our interests.' It could be argued that it was therefore in Britain's imperial interests to ensure the Sinhalese dominated Ceylon's government after independence, to prevent any pan-Tamil solidarity that might expand India's territorial influence and give her the capacity to become a superpower.

23. Ponnambalam, *Sri Lanka*, p. 105.
24. Ponnambalam, *Sri Lanka*, p. 113.
25. Ponnambalam, *Sri Lanka*, p. 131.
26. Ponnambalam, *Sri Lanka*, p. 180.
27. Ponnambalam, *Sri Lanka*, p. 183.
28. Immediately after the JVP uprising, Ceylon's police had asked Britain for more riot-control equipment, including tear gas guns, riot shields, hand grenades, flares, automatic pistols, centrifugal water pumps, water bowsers and body armour.
29. Nanise Loanakadavu, 'Special Breed of Men Who Fought Secret Wars', *Fiji Times*, 29 April 2013.
30. Anita Pratap, *Sunday Magazine*, 11-17 March 1984.
31. Interview with David Gladstone, Wotton House, November 2017.
32. John Percival Morton, *Indian Episode*, unpublished memoir presented to the British Library in 1982, pp. 5–6.
33. AIR 8/2803, 'Annual and End of Tour Report', Defence Adviser to the British high commissioner in Sri Lanka, August 1978–November 1979, p. 8, Lieutenant Colonel K. H. J. Reynolds.
34. MI5, 'Your Enquiry', email from noreply@mi5.gov.uk, 3 July 2018.

35. FSC 382/1, 'UK Assistance to Sri Lankan Police', 1983. The Morton report is referred to in a paper marked 'Sri Lanka Police – Visit by Overseas Police Adviser to Colombo 7–11 March 1983'.

36. Rusty Firmin, *The Regiment: 15 Years in the SAS*, Osprey Publishing, Oxford, 2016, p. 161.

37. PREM 19/1395, 'Sri Lanka: Visit of Foreign Minister', 21 December 1981.

38. PREM 19/1395, 'Sri Lanka: Visit of Foreign Minister', 21 December 1981.

39. PREM 19/1395, 'Sri Lanka: Visit of Foreign Minister', 21 December 1981.

40. PREM 19/1395, 'Tamil Eelam', 18 December 1981, and 'Sri Lanka: Visit of Foreign Minister', 17 December 1981.

Chapter 4

1. FCO 8/3336, 'Annual Report for 1978', British defence attaché in Muscat.

2. FCO 8/3336, 'Annual Report for 1978', British defence attaché in Muscat.

3. NAM, Colonel Dim Robbins OBE MC, deputy director general, Oman MOD, 30 April 1978.

4. FCO 37/3977, 'KMS Ltd', 29 November 1985. See also Alastair MacKenzie, *Pilgrim Days*, Osprey, Oxford, 2019, chapter 7.

5. Hart-Davis, *The War That Never Was*, p. 12.

6. Cabinet, Joint Action Committee, J.A. (64)3, 'The MI6 Role and Relationships with Departments of State and the Armed Services in the Conduct of Deniable Operations in Conditions Short of War', 1 October 1964.

7. Hart-Davis, *The War That Never Was*, p. 19.

8. On 9 May 1979 at 7.35 am. FCO 8/3546, 'Annual Report for 1979', British defence attaché in Muscat.

9. John Maslin, 'Family Donate Medal to Collegiate', *Wanganui Chronicle*, 12 March 2015.

10. FCO 8/3546, HM Ambassador Muscat to Lord Carrington, 19 February 1980.

11. NAM, Landon to Templer, 6 February 1979.

12. FCO 8/3954, 'Annual Report for 1980', British defence attaché in Muscat.

13. FCO 8/3954, 'The Military Scene', I. T. M. Lucas to R. E. Palmer, 27 October 1981: 'Creasey may be moving towards grasping the nettle of trying to bring the Royal Guard, ROP, SSF *et al* (underline et al) under some central control, at least for the purposes of expenditure. If he succeeds – and the difficulties are not to be underestimated – this would represent a significant shift in the balance of power in the country'.
14. Army number 484596.
15. Companies House, 'Saladin Security Annual Return', 1979.
16. FCO 37/3977, F43, 'Private Security Organisations', December 1984.
17. Army number 489891.
18. Dix Noonan Webb Ltd, 'A Good Series of Awards to Members of the SAS', 16 December 2003, www.dnw.co.uk/auction-archive/special-collections/lot.php?specialcollection_id=291&special collectionpart_id=287&lot_id=95117 (accessed 6 June 2019).
19. Alistair Kerr, *Betrayal: The Murder of Robert Nairac GC*, Liverpool Academic Press, Liverpool, 2015.
20. Kerr, *Betrayal*.
21. FCO 37/3977, 'KMS Ltd', 29 November 1985.
22. FCO 8/3954, 'Oman: Brigadier Landon', J. C. Moberly to Miers, 24 November 1981.
23. FCO 8/3954, 'Oman: Brigadier Landon', J. C. Moberly to Miers, 24 November 1981.
24. FCO 8/5511, 'Defence Attaché Report for 1983', 30 July 1984.
25. European Commission historic archives in Brussels, BAC 163/1991 655, 1979–82.
26. Surrey History Centre, 'Elmbridge Borough Council, Councillors, Wards, Committees, Dates of Meetings', 1982–3.
27. Surrey History Centre, 7723/5/1/10, p. 23.
28. FCO 7/4804, 'Diplomatic Representation of the UK in Uruguay', 1982.
29. Surrey History Centre, 7723/5/1/10, p. 158.
30. Surrey History Centre, 7723/5/1/10, p. 305.
31. Surrey History Centre, 7723/5/1/10, 23 June 1983, p. 212.
32. FSC 382/1, 'Aid for Sri Lanka Police', from British High Commission (BHC) to South Asian Department, 11 February 1983.
33. FCO 37/3130, 'Internal Situation – Sri Lanka', Lieutenant Colonel Sale to MOD UK, 7 February 1983.

34. FSC 382/1, 'Sri Lanka Police – Visit by Overseas Police Adviser to Colombo 7–11 March 1983', Robert P. Bryan, 17 March 1983.
35. FSC 382/1, 'Sri Lanka Police – Visit by Overseas Police Adviser to Colombo 7–11 March 1983', Robert P. Bryan, 17 March 1983.
36. FSC 382/1, 'Sri Lanka Police – Visit by Overseas Police Adviser to Colombo 7–11 March 1983', Robert P. Bryan, 17 March 1983.
37. FSC 382/1, 'Sri Lanka Police – Visit by Overseas Police Adviser to Colombo 7–11 March 1983', Robert P. Bryan, 17 March 1983.
38. FSC 382/1, 'Sri Lanka Police – Visit by Overseas Police Adviser to Colombo 7–11 March 1983', Robert P. Bryan, 17 March 1983.
39. FSC 382/1, 'Overseas training for Sri Lanka Police Officers', Sri Lankan High Commission London to Peel Centre, UK Police Training College, 15 April 1983.
40. FSC 382/1, 'Sri Lankan Police Training', FCO South Asian Department to the Overseas Police Advisor, and 'Para-military Training Course in the UK', from R. Sunderalingham (Senior Deputy IGP), 3 May 1983.
41. Ponnambalam, *Sri Lanka*, p. 223.
42. Ponnambalam, *Sri Lanka*, p. 224.
43. Senior Deputy Inspector-General Herbert W. H. Weerasinghe, from Sri Lanka's CID, and Assistant Superintendent K. S. Padiwita, from the Intelligence Services Division (i.e. Special Branch).
44. FSC 382/1, 'International Conference on Terrorist Devices and Methods, Deep Cut – June 1983', R. Rajasingham (inspector general of police) to Robert P. Bryan, 15 March 1983.
45. FSC 382/1, 'Sri Lanka Police', Robert P. Bryan to Paul Buxton (Northern Ireland Office), 25 March 1983.
46. Ponnambalam, *Sri Lanka*, pp. 225–6.
47. FCO 37/3131, 'Internal Security', Lieutenant Colonel Sale to MOD UK, 22 September 1983.
48. FCO 37/3131, 'Internal Security', Lieutenant Colonel Sale to MOD UK, 22 September 1983.
49. Anthony Mascarenhas, 'Terrorists Face Big Crackdown', *Sunday Times*, 28 August 1983.
50. FCO 37/3150, 'Sri Lanka: Security Situation', Sir J. Nicholas to FCO London, 6 September 1983.
51. FCO 37/3150, 'Sri Lanka: Security Situation', Sir J. Nicholas to FCO London, 6 September 1983.

52. FCO 37/3150, 'Sri Lanka: Security Situation', 9 September 1983, and 'Sri Lanka: Request for Anti-terrorist Training', 12 September 1983.
53. FCO 37/3150, 'Sri Lanka: Request for Military Training', South Asia Department to Ministry of Defence DS11, 9 September 1983.
54. FCO 37/3150, 'Sri Lanka – Request for Military Assistance', MOD DS11 to FCO SAD, 21 September 1983.
55. FCO 37/3150, 'Basic and Advance Courses in Special Branch Work in the U.K.', Merril Gunartne to Lieutenant Colonel E. G. R. Sale, 19 September 1983. Gunartne also attended MI5 courses in Bond Street.
56. FCO 8/2006, Antony Parsons to Le Quesne, 12 February 1973, cited in Takriti, *Monsoon Revolution*, p. 287.
57. FCO 37/3150, 'Sri Lankan Request for Military Training', 20 September 1983.
58. FCO 65/1768, 6 August 1976, West African Department, *Mercenaries: The Recommendations of the Diplock Committee.*
59. FCO 37/3150, 'Sri Lanka Request for Anti Terrorist Training', 26 September 1983, FCO London to BHC Colombo.
60. FCO 37/3150, 'Anti Terrorist Training', undated.
61. FCO 37/3150, Clutterbuck to FCO South Asia Department, 19 October 1983.
62. FCO 37/3150, 'Sri Lanka: Training in Counter Terrorism', 28 October 1983.
63. His firm already had some familiarity with Sri Lanka, having advised their flag carrier Air Lanka on airport security.
64. FCO 37/3150, 'Sri Lanka: Armed Forces Training', minutes of meeting between Clutterbuck, API and Burton (SAD), 19 October 1983.
65. FCO 37/3150, 'Anti Terrorist Training', undated.
66. FCO 37/3150, 'Anti Terrorist Training', BHC Colombo to FCO London, 4 November 1983.
67. FCO 37/3150, 'Anti Terrorist Training', BHC Colombo to FCO London, 9 November 1983.
68. FCO 37/3150, 'Anti Terrorist Training', BHC Colombo to FCO London, 15 November 1983.
69. FCO 37/3132, 'Discussion with General Attygalle', Lieutenant Colonel Sale to BHC Colombo, 6 December 1983.

70. FCO 37/3132, 'Sri Lanka: Internal Security', J. P. P. Nason to FCO SAD, 16 December 1983.
71. FCO 37/3151, 'Supply of Riot Control Equipment to Sri Lanka', FCO SAD to Lady Young, private secretary, 10 August 1983.
72. FCO 37/3151, 'Supply of Armoured Cars to Sri Lanka', FCO SAD to Lady Young, private secretary, 3 October 1983.

Chapter 5

1. John Smith, 'Impressions of Nicaragua', *Free Nicaragua*, April 1980.
2. PREM 19/2367, J. S. Wall (FCO) to B. G. Cartledge (Downing Street), 26 July 1979. For Thatcher's reluctance, see reply sent 28 July 1979. Thatcher allowed recognition on 1 August 1979.
3. Parliamentary Human Rights Group, 'Report of a British Parliamentary Delegation to Nicaragua to Observe the Presidential and National Assembly elections', 4 November 1984.
4. Smith, 'Impressions of Nicaragua'.
5. PLOTE, *Spark*, vol. 1, no. 2, January 1985, Madras, India, pp. 27–30.
6. 'Assistance to the Nicaraguan Resistance', Oliver North to Robert McFarlane, 4 December 1984, declassified in the *Report of the Congressional Committees Investigating the Iran– Contra Affair: With Supplemental, Minority, and Additional Views*, 1987, Appendix A, vol. 1, pp. 901–5.
7. *World in Action*, 'Uncle Sam's British Mercenary', Granada TV, 25 July 1988. See also Nick Davies, 'The Assassination Business', *Scotsman*, 26 July 1988.
8. Title redacted in part, Oliver North to Robert McFarlane, 20 December 1984, declassified in the *Report of the Congressional Committees Investigating the Iran–Contra Affair: With Supplemental, Minority, and Additional Views*, 1987, Appendix A, vol. 1, pp. 264–6.
9. FCO 99/2891, Acland in Washington to London, 9 July 1988.
10. Reagan Library, Box 90902, 'Meeting with British Prime Minister Margaret Thatcher Margaret', 22 December 1984.
11. FCO 99/2891, 'Jocaks – Commentary on Documents to be Released', 13 July 1988.
12. Title redacted, Oliver North to Adolfo Calero, undated, declassified in the *Report of the Congressional Committees Investigating the*

Iran–Contra Affair: With Supplemental, Minority, and Additional Views, 1987, Appendix A, vol. 1, pp. 222–4.

13. Imperial War Museum, A 26916.

14. Surrey History Centre, 7723/5/1/11, p. 1095.

15. He named Peter Slade and Nobby Clarke as KMS members.

16. Nick Davies, 'Marxist Revolutionaries Who Joined the Irish Republican Cause', *The Observer*, 9 February 1986.

17. Supplement, *London Gazette*, 22 June 1965.

18. FCO 37/3509, 'Sri Lanka – Internal Security', Defence Adviser (Sale) to MOD London, 31 August 1984.

19. FCO 37/3509, 'Sri Lanka – Internal Security', Defence Adviser (Sale) to MOD London, 31 August 1984.

20. *The Hindu*, Madras, 16 October 1984.

21. FCO 37/3509, Untitled, Lieutenant Colonel Sale to MOD UK, 3 September 1984.

22. 'Sri Lanka Army Terrorises Another Jaffna School', *Tamil Information Monthly*, 15 November 1984.

23. FCO 37/3509, F170, 20 September 1984, Defence Attaché to MOD.

24. FCO 37/3509, 'Sri Lanka – Internal Security', Defence Adviser (Sale) to MOD London, 5 October 1984.

25. Surrey History Centre, 7723/5/1/12, p. 456.

26. FCO 37/3131, 'Internal Security', Lieutenant Colonel Sale to MOD UK, 22 September 1983.

27. FCO 37/3538, 'General Attygalle: Visit to UK', 26 September 1984, BHC Colombo to FCO London.

28. FCO 37/3538, 'Visit by Sri Lankan Defence Secretary', 10 October 1984, FCO SAD to PS/Lady Young.

29. FCO 37/3538, Letter from General Attygalle to Sir John Hermon, 23 October 1984.

30. FCO 37/3548, G. W. Harding to Mr Cleghorn, 28 December 1984.

31. FCO 37/3977 was retained by the Foreign Office and FCO 37/3978 is still retained by the National Archives.

32. FCO 37/3977, F3, 2 January 1985, C. R. Budd to T. C. Wood.

33. FCO 37/3977, F4, 17 January 1985, FCO London to BHC Colombo, secret dedip ('dedip' is a security classification used by diplomats to signify that a telegram is highly sensitive).

34. FCO 37/3977, F5, 'Sri Lanka: British Involvement', 21 January 1985, T. C. Wood to Howe, Lady Young, Harding and Permanent Under Secretary's Department.

35. FCO 37/3977, F6, 'Sri Lanka: British Involvement', 22 January 1985, P. F. Ricketts to T. C. Wood, Bruce Cleghorn.
36. The ambassador seemed keen to sell the Sri Lankan forces British guns even if he did not want them to receive British training.
37. FCO 37/3977, F10, 'KMS Limited', UK High Commissioner to Sir W. Harding, 28 January 1985.
38. FCO 37/3977, F12, 'KMS Limited', UK High Commissioner to Sir W. Harding, 29 January 1985.
39. FCO 37/3977, F13, 'Answers to the SoS's Questions', Mrs A. Glover, 1 February 1985.
40. FCO 37/3977, F15, 'Involvement of KMS Ltd in Sri Lanka', Bruce Cleghorn, 4 February 1985.
41. Written on FCO 37/3977, F15, 'Involvement of KMS Ltd in Sri Lanka', dated 5 February 1985.
42. FCO 37/3977, F19, 'KMS Limited', UK High Commissioner to Sir W. Harding, 7 February 1985.
43. Interviewed by Phil Miller on 4 December 2017 in Bordeaux.

Chapter 6

1. *World in Action*, 'Uncle Sam's British Mercenary', Granada TV, 25 July 1988.
2. Alvin Shuster and Juan M. Vasquez, 'Explosions Hit Army Hospital in Managua', *LA Times*, 7 March 1985.
3. Tim Golden, 'Explosions Rock Managua, Damage Military Hospital', *United Press International*, 7 March 1985.
4. FCO 99/2385, 'Press Release on the State of National Emergency Decreed in Nicaragua on 15 October 1985'.
5. US Congress Iran–Contra Investigation, 14 May 1987.
6. *World in Action*, 25 July 1988.
7. US National Security Council, 'Timing and the Nicaraguan Resistance Vote', Oliver North to Robert McFarlane, 20 March 1985.
8. *World in Action*, 25 July 1988.
9. Surrey History Centre, 7723/5/1/12, p. 1255.
10. FCO 99/2621, 'Blowpipe: Application for Sale to Honduras', M. L. Croll to Private Secretary on 5 September 1985, and addendum by David Thomas on 6 September 1985.
11. FCO 37/3928, 'Batticaloa District', Alasdair MacDermott, 11 March 1985.

12. FCO 37/3973, 'Security Situation', BHC Colombo to FCO London, 18 January 1985.

13. FCO 37/3973, 'Incidents Week Ending 15 Feb 85', Lieutenant Colonel Holworthy to MOD UK.

14. FCO 37/3973, 'Sri Lanka: Security Situation', BHC Colombo to FCO London, 8 March 1985. The EPRLF at that stage had the respect of Ambalavaner Sivanandan, a Tamil Marxist intellectual who was director of the Institute of Race Relations in London.

15. PLOTE, *Spark*, January1985.

16. FCO 37/3973, 'Batticaloa – The Next Jaffna/Mannar', Alasdair MacDermott, 4 March 1985.

17. FCO 37/3973, 'Visit to Batticaloa', Justin Nason, 8 March 1985.

18. FCO 37/3928, 'Sri Lanka Internal – The Disputed Areas Jaffna, Mannar, Batticaloa', Alasdair MacDermott, 15 March 1985.

19. Batticaloa Citizens Committee, 'Arrests Made at Koddaikallar on 24 January 1985', copy found in FCO 37/3928.

20. Batticaloa Citizens Committee, 'Thirukovil Area, 27 January 1985', copy found in FCO 37/3928.

21. *Tamil Guardian*, 'Remembering the Batticaloa Lake Road Massacre of 1985', 13 November 2018.

22. FCO 37/4336, 'Report on the Visit of the Committee for the Monitoring of Cessation of Hostilities to Batticaloa on Wednesday 13th and Thursday 14th November 1985'.

23. Interview with Father John Joseph Mary in Batticaloa, August 2017.

24. FCO 37/3977, F20, 'Sri Lanka: KMS Ltd', Sir W. Harding to Terence Wood FCO SAD, 15 March 1985.

25. FCO 37/4012, 'Visit by Sir William Harding, Deputy Under-Secretary of State for Foreign and Commonwealth Affairs, to Sri Lanka', March 1985.

26. FCO 37/4012, 'Visit by Sir William Harding, Deputy Under-Secretary of State for Foreign and Commonwealth Affairs, to Sri Lanka', March 1985.

27. FCO 37/4012, 'Meeting between Sir William Harding and General Attygalle (Secretary, MOD)', 26 March 1985.

28. FCO 37/3977, F25, 'Sri Lanka: KMS Ltd', Sir W. Harding to Terence Wood, 10 April 1985.

29. FCO 37/3977, F24, 'KMS Limited', John Stewart to Terence Wood, 13 May 1985. Stewart told President Jayewardene that anti-guerrilla training required more than six months to succeed.

30. FCO 37/3979, F11, 'Sri Lanka', R. J. O'Neill to Dr Wilson. Draws parallel with British training of Mugabe's forces in Zimbabwe 'despite the tribal tensions ... and the allegations of abuse'.

31. Athithan Jayapalan, 'Eelam, Wahhabism and Sri Lankan COIN; the Case of Tamils and Muslims', *TamilNet,* 25 April 2019.

32. PLOTE, *Spark,* January 1985.

33. Mark P. Whitaker, *Learning Politics from Sivaram,* Pluto Press, London, 2006, p. 99.

34. LTTE publication, 1985. Copy contained in FCO 37/4336, F30.

35. FCO 37/3929, F96, Holworthy to MOD/FCO, 22 July 1985.

36. FCO 37/3973, F72, BHC to London 17 April 1985. See also 23 April 1985: 'The destruction of property seems to have been very great. The STF are being blamed for failing to prevent the escalation of violence.'

37. FCO 37/3974, Holworthy, 10 May 1985.

38. FCO 37/4009, 'Inter-communal Violence between Muslim and Tamils', 14 April 1985.

39. FCO 37/4009, Holworthy to MOD, 29 April 1985.

40. FCO 37/4009, BHC to FCO London, 9 May 1985.

41. FCO 37/3979, F15, Terence Wood (SAD) to Dr Wilson, 26 April 1985.

42. FCO 37/4009, 25 April 1985.

43. FCO 37/3929, F96, Holworthy to MOD/FCO, 22 July 1985.

44. FCO 37/3929, F96, Holworthy to MOD/FCO, 22 July 1985.

45. FCO 37/3929, F43, BHC to London, 3 May 1985.

46. Shyam Batia, *The Observer,* 14 April 1985. He also interviewed Manendra Kesivapillai, 23, a second-year medical science student from Jaffna university who was tortured by the STF after being snatched in Batticaloa in January 1984.

47. FCO 37/3975, 'MacDermott's Visit to Mannar on 26–28 April 1985', 30 April 1985. I do not know the full extent of their atrocities in Mannar, but in 2018 a significant mass grave was unearthed there, across the road from a police camp. Its precise origins are still the subject of considerable debate.

48. FCO 37/4336, 'Sri Lanka: Internal Security', BHC Colombo to FCO London, 24 February 1986.

49. FCO 37/4336, 'Sri Lanka: Internal Security', BHC Colombo to FCO London, 24 February 1986.
50. FCO 37/4336, 'Sri Lanka: Internal Security', BHC Colombo to FCO London, 3 March 1986.
51. FCO 37/3975, MacDermott to London, 8 August 1985.
52. FCO 37/3975, 'Joint Operations Council', Alasdair MacDermott to FCO SAD, 8 August 1985.
53. FCO 37/3930, 'Sri Lanka – Internal Situation', Lieutenant Colonel Holworthy to FCO SAD, 19 August 1985.
54. Surrey History Centre, 7723/5/1/12, 4 September 1985, p. 610.

Chapter 7

1. Interview with Phil Miller, Prague, 6 November 2017.
2. Robin Horsfall, *Fighting Scared*, Cassell, London 2002.
3. This could have been on 19 February 1986, near Kantale, Trincomalee, where between 19 and 35 Sinhalese villagers were killed. In FCO 37/4336, F36, there is a report of a landmine exploding on 19 February in the Sinhalese settlement area at Seruvawila: 'A convoy of colonists was being escorted by the army to the market at Kantalai.' It said 35 civilians and four soldiers were killed.
4. Sri Lankan Army, 'Maduru Oya ATS Reaches 33 Years of Age', 19 January 2018, www.army.lk/news/maduru-oya-ats-reaches-33-years-age (accessed 6 June 2019).
5. Jonathan Humphries, 'Former Army Captain Lived the High Life After Duping the Taxman', *Liverpool Echo*, 15 February 2016.
6. FCO 37/4299, 'Afghanistan: Training for Resistance from KMS', J. Poston (South Asia Department) to Mr Dewar and Mr Wood, 19 May 1986.
7. FCO 37/4299, 'Afghanistan: Training for Resistance from KMS', J. Poston (South Asia Department) to Mr Dewar and Mr Wood, 19 May 1986.
8. FCO 37/2215, 'Afghanistan', R. D. Lavers to Head of Chancery, 24 January 1980. See also FCO 37/2215, 'Afghanistan', 4 March 1980.
9. FCO 37/2215, Lord Carrington to Lord Gore-Booth, 15 February 1980.
10. FCO 37/2215, 'Afghanistan', K.B.A. Scott to Peter M. Maxey, Cabinet Office, 2 June 1980.

11. FCO 37/2215, 'Afghanistan: Supporting the Resistance', Secretary of State to Mr Cane, undated draft.
12. Steve Coll, *Ghost Wars*, Penguin, New York, 2004, p. 123.
13. George Cale, *My Enemy's Enemy, the Story of the Largest Covert Operation in History: The Arming of the Mujahideen by the CIA*, Atlantic Books, London, 2003, pp. 197–201.
14. Cale, *My Enemy's Enemy*, pp. 197–201.
15. Coll, *Ghost Wars*, p. 135.
16. Email from Prisoner Location Service, 10 October 2018.
17. *London Gazette*, 12 June 1982. Lists him as Timothy Andrew Smith, number 23662211, Warrant Officer Class 2.
18. Tim Smith, *The Reluctant Mercenary: The Recollections of a British Ex-Army Helicopter Pilot in the Anti Terrorist War in Sri Lanka*, Book Guild, Lewes, 2002, p. 11.
19. Smith, *The Reluctant Mercenary*, p. 13.
20. Smith, *The Reluctant Mercenary*, p. 14.
21. Smith, *The Reluctant Mercenary*, pp. 16–17.
22. NESOHR, 'Lest We Forget, Massacres of Tamils 1956–2001, part 1', 2007, p. 96.
23. Smith, *The Reluctant Mercenary*, p. 19.
24. Smith, *The Reluctant Mercenary*, p. 21.
25. Smith, *The Reluctant Mercenary*, p. 22.

Chapter 8

1. The Tamil Eelam Liberation Organization.
2. FCO 37/3930, 'Sri Lanka', Sir Robert Wade-Gery in New Delhi to BHC Colombo, 23 September 1985.
3. FCO 37/3977, F26, John Stewart to MOD UK, 18 June 1985.
4. FCO 37/3977, F27, FCO London to John Stewart, 4 July 1985.
5. FCO 37/3977, F29, John Stewart to MOD UK, 10 July 1985.
6. FCO 37/3977, F29, John Stewart to MOD UK, 10 July 1985.
7. FCO 37/3977, F36, Peter Ricketts to Terence Wood, 15 July 1985.
8. FCO 37/3977, F29, Terence Wood and then Dr Wilson to Sir W. Harding, 12 July 1985.
9. FCO 37/3977, F32, FCO London to BHC Colombo, 17 July 1985.
10. FCO 37/3929, 'Sri Lanka: Inter-communal Problem and Samanalawewa Power Station Project', Bruce Cleghorn to Lady Young's private secretary, 19 July 1985.

11. FCO 37/3929, 'KMS Limited', BHC Colombo to FCO London, 19 July 1985.

12. FCO 37/3929, 'Sri Lanka', FCO London to BHC Colombo, 22 July 1985.

13. FCO 37/3929, 'Sri Lanka: Inter-communal Problem and Samanalawewa Power Station Project', Bruce Cleghorn to Lady Young's private secretary, 19 July 1985.

14. FCO 37/3930, 'Sri Lanka – Internal Situation', 19 August 1985, Lieutenant Colonel Holworthy to FCO SAD.

15. FCO 37/3930, F122, BHC to London, 27 August 1985. Stewart (BHC) visited security forces in Jaffna on 21–3 August 1985.

16. FCO 37/3977, F40, 'KMS in Sri Lanka', John Stewart to Terence Wood, 5 August 1985.

17. FCO 37/3930, F148, BHC to London, 16 September 1985.

18. FCO 37/3930, F222, 12 November 1985.

19. FCO 37/3930, F218, Alasdair MacDermott to Cabinet Office, 20 November 1985.

20. FCO 37/3977, F50, 'Private Security Organisations', 29 October 1985.

21. FCO 37/3977, F52, John Stewart to FCO London, 29 November 1985.

22. FCO 37/3977, F53, John Stewart to FCO London, 4 December 1985.

23. FCO 37/3977, F54, John Stewart to FCO London, 6 November 1985.

24. FCO 37/3977, F55, 'KMS Ltd', J. D. K. Andrews (CICO/DI ROW) to FCO SAD, 29 November 1985.

25. FCO 37/4347, F1, 'Summary of Call on Lady Young by Mr Amir-thalingam, 11 December 1985 at 11.10 am'.

26. FCO 37/3930, F214, FCO SAD to Permanent Under Secretary's Department, 3 December 1985.

27. FCO 37/4336, Untitled, High Commissioner John Stewart to Terrance Wood (SAD), 22 January 1986.

28. FCO 37/4336, Untitled, High Commissioner John Stewart to Terrance Wood (SAD), 22 January 1986.

29. FCO 37/4336, 'Sri Lanka Internal Security', BHC Colombo to FCO London, 24 February 1986.

30. FCO 37/4354, with F5, 'India/Sri Lanka', Wade-Gery to FCO London, 25 February 1986.

311

31. FCO 37/4354, F5, 'Sri Lanka: Role of KMS Limited', Terence Wood SAD, 27 February 1986.
32. FCO 37/4336, 'Sri Lanka: Internal security', 3 March 1986.
33. FCO 37/4336, 'Sri Lanka: Internal Security', BHC Colombo to FCO London, 10 March 1986.
34. FCO 37/4336, 'Sri Lankan Situ: Indian Mediation', Canadian embassy in Delhi, 28 February 1986.
35. FCO 37/4336, 'Sri Lanka: Reports by Guardian Correspondent', BHC Colombo to FCO London, 6 March 1986.
36. FCO 37/4336, 'Sri Lanka: KMS', Sir Wade-Gery in Delhi to FCO London, 10 March 1986.
37. FCO 37/4336, 'Secretary of State's Meeting with Indian Foreign Minister', Sir Wade-Gery in New Delhi, 1 April 1986.
38. FCO 37/4354, F18, 'Sri Lanka: KMS', John Stewart to FCO London, 21 March 1986.
39. FCO 37/4336, 'Secretary of State's Meeting with Indian Foreign Minister', Sir Wade-Gery in New Delhi, 1 April 1986.
40. FCO 37/4336, 'Discussion with the Indian High Commissioner', John Stewart internal memo, 14 March 1986.
41. FCO 37/4354, F18, 'Sri Lanka: KMS', John Stewart to FCO London, 21 March 1986.
42. FCO 37/4336, Untitled, High Commissioner John Stewart to Terrence Wood (SAD), 19 March 1986.
43. FCO 37/4336, 'Sri Lanka Sitrep', Howe to BHC Colombo, 7 April 1986.
44. FCO 37/4336, 'Sri Lanka: Internal Situation', High Commissioner Stewart to FCO London, 9 April 1986.
45. FCO 37/4336, 'Air, Ground and Naval Attack on VVT on 12 March 1986', VVT Citizens Committee.
46. FCO 37/4354, F12, 'Activities of British Subjects in Sri Lanka', E. Denza to Mr Hook (SAD), 13 March 1986.
47. FCO 37/4354, F13, FCO London to BHC Colombo, 17 March 1986.
48. FCO 37/4354, F17, 'KMS in Sri Lanka', Sherard Cowper-Coles to Terrence Wood (SAD), 20 March 1986.
49. FCO 37/4354, F20, 'Sri Lanka: KMS Activities', Terrence Wood (SAD) to Lady Young's private secretary, 24 March 1986.
50. FCO 37/4354, F23, 'India/Sri Lanka: KMS', FCO London to BHC Colombo, 27 March 1986.

51. Surrey History Centre, 7723/5/1/14, 22 April 1986, p. 1623.

52. Smith, *The Reluctant Mercenary*, p. 24.

53. Hansard, 22 May 1986.

54. Smith, *The Reluctant Mercenary*, p. 32.

55. Smith, *The Reluctant Mercenary*, p. 34.

56. Smith, *The Reluctant Mercenary*, p. 46.

57. Smith, *The Reluctant Mercenary*, p. 51.

58. FCO 37/4354, F53, dated slightly earlier to 29 May 1986.

59. Smith, *The Reluctant Mercenary*, p. 51.

60. Smith, *The Reluctant Mercenary*, p. 55.

61. FCO 37/4354, F53.

62. Smith, *The Reluctant Mercenary*, p. 57.

63. Smith, *The Reluctant Mercenary*, pp. 58-9.

64. FCO 37/4354, F53.

65. US Congress Iran–Contra Investigation, testimony of Robert Dutton, 27 May 1987.

66. FCO 99/2385, 'Nicaragua', 3 June 1986, David Thomas to Lady Young's private secratary. Thomas refers to a meeting on June 2, 1986, between 'Elliot Abrams and other senior American officials in talks chaired by Sir W Harding'.

67. Dana Walker and Anne Saker, *United Press International*, 27 May 1987.

68. *Report of the Congressional Committees Investigating the Iran–Contra Affair: With Supplemental, Minority, and Additional Views*, 1987, Appendix A, vol. 1, p. 1214.

69. *World in Action*, 25 July 1988.

70. FCO 99/2620, 'Background for PMQ's', 26 March 1987.

71. FCO 99/2621, 'Clover Leaf V: Procurement of Blowpipe for the Contras', David Thomas to Sir William Harding, 22 May 1986.

72. FCO 99/2621, 'Clover Leaf V: Procurement of Blowpipe for the Contras', David Thomas to Sir William Harding, 22 May 1986. See also FCO 99/2405, 'Arms Supplies to Chile', Howe to British Ambassador Santiago, 30 May 1986.

73. President's Special Review Board (Tower Commission), 26 February 1987.

74. FCO 99/2891, Acland in Washington to London, 9 July 1988.

75. FCO 99/2620, 'Prime Minister's Question Time, Confidential Background on Supply of Blowpipe to Contras', 26 March 1987.

Chapter 9

1. Smith, *The Reluctant Mercenary*, p. 65.
2. Smith, *The Reluctant Mercenary*, p. 65.
3. Smith, *The Reluctant Mercenary*, p. 68.
4. FCO 37/4354, F35, 'Sri Lanka: KMS', Richard J Codrington to Head of Chancery.
5. Smith, *The Reluctant Mercenary*, p. 72. He names the pilots as Don Burton and 'Stan from Alderney'.
6. FCO 37/4338, F246, 'Sri Lanka: The Armed Forces', J. H. Jones, 3 October 1986.
7. FCO 37/4354, F40, 'KMS', BHC Colombo to FCO London, 16 July 1986.
8. FSC 074/1, F30, 'Tour of the Eastern Province 9–10 July 1986', Lieutenant Colonel Holworthy, 11 July 1986.
9. FCO 37/4354, F41, 'Sri Lanka: KMS', 17 July 1986.
10. FCO 37/4354, F42, 'Sri Lanka: KMS', 18 July 1986.
11. FCO 37/4354, F48, 'Sri Lanka: KMS', R. N. Culshaw to Terence Wood (SAD), 21 July 1986.
12. FCO 37/4354, F43, 'Sri Lanka: KMS', E. Denza to Terence Wood (SAD), 22 July 1986.
13. FCO 37/4354, F49, 'KMS: The Mankeni [*sic*] attack', Richard J. Langridge to N. Hook (SAD), 23 July 1986.
14. FCO 37/4354, F50, 'Sri Lanka: KMS', Terence Wood (SAD) to Sir W. Harding, 6 August 1986.
15. FCO 37/4354, F51, 'KMS – Use by HMG', author unclear, 11 August 1986. See also F58, undated, from H. J. Nicholls (Security Department). There is also the European Commission Historical Archive, *Ouganda*, BAC 163/1991655, 1979–82.
16. FCO 37/4354, F55, 'Record of a Conversation Between Sir William Harding, Deputy Under Secretary of State, and Colonel James Johnson, Chairman of KMS, on 8 August 1986 at 11.30 am'.
17. FCO 37/4347, F38, 'Call on Lady Young by Lalith Athulathmudali', 11 August 1986.
18. FCO 37/4354, F61, 'Sri Lanka: KMS, 4 September 1986', J. Poston (SAD).
19. FCO 37/4354, F63, 'KMS', 11 September 1986, John Stewart to London.
20. Smith, *The Reluctant Mercenary*, pp. 110–11.

21. FCO 37/4338, F231, 29 September 1986, John Stewart to FCO London.
22. Smith, *The Reluctant Mercenary*, p. 147.
23. Smith, *The Reluctant Mercenary*, p. 177.
24. Smith, *The Reluctant Mercenary*, p. 217.
25. Smith, *The Reluctant Mercenary*, p. 252.
26. Smith, *The Reluctant Mercenary*, pp. 245-6.
27. FCO 37/4338, F250, 'Death of Deutsche Welle engineer', J. H. Jones, 13 October 1986.
28. *Die Welt*, 27–8 September 1986. See also Reuters, 'Sri Lanka Extends Measures', *New York Times,* 28 September 1986.
29. Aruguam Bay Information, 'DW', 20 November 2006, www.arugam. info/2006/11/20/deutsche-dauer-welle-dw/ (accessed 6 June 2019).
30. FCO 37/4338, F240, 'Death of Deutsche Welle engineer', J. H. Jones, 2 October 1986.
31. US Select Committee on Secret Military Assistance to Iran and the Nicaraguan Opposition, testimony of Shirley A. Napier, pp. 269–70, March–April 1987.
32. US Select Committee on Secret Military Assistance to Iran and the Nicaraguan Opposition, testimony of Shirley A Napier, p. 327, March–April 1987.
33. Smith, *The Reluctant Mercenary*, pp. 258-9.
34. PREM 19/2875, 20 May 1987.
35. Jeremy Corbyn, speech to Tamils of Lanka exhibition, Tolworth, 19 May 2019.
36. I am grateful to Felix Bazalgette, who wrote about this important episode of Tamil history for *New Internationalist* magazine, no. 516, November–December 2018, in an excellent feature titled 'Between the Devil and the Deep Blue Sea'.
37. FCO 37/4705, F12, 'KMS', FCO London to BHC Colombo, 5 February 1987.
38. FCO 37/4705, F13, 'Sri Lanka: KMS', R. A. Burns, 17 February 1987.
39. FCO 37/4705, F15, BHC Colombo to FCO London, 23 February 1987.
40. FCO 37/4705, F17, 'KMS: Possible Withdrawal from Sri Lanka', FCO London to BHC Colombo, 4 March 1987.
41. FCO 37/4705, F18, 'KMS', BHC Colombo to FCO London, 5 March 1987.

42. FSC 074/1, F39, 'Meetings with DNIC and IGP', Lieutenant Colonel Holworthy, 29 October 1986.
43. FCO 37/4705, F19, 'KMS', BHC Colombo to FCO London, 6 March 1987.
44. FCO 37/4705, F21, 'KMS: Sri Lanka', R. N. Culshaw, 4 March 1987.
45. FCO 37/4705, F28, 'KMS', John Stewart to FCO London, 9 March 1987.
46. FCO 37/4705, F30, 'KMS', D. H. Gillmore to Colonel Johnson, 9 March 1987.
47. FCO 37/4706, F40, David Walker to D. H. Gillmore, 23 March 1987.
48. FCO 37/4706, F43, R. A. Burns to Private Secretary of Lady Young, 24 March 1987.
49. FCO 37/4706, F 44, FCO London to BHC Colombo, 31 March 1987.

Chapter 10

1. US Congress Iran–Contra Investigation, testimony of Adolfo Calero, 20 May 1987.
2. Phil Davison, 'Adolfo Calero', *The Independent*, 5 June 2012.
3. US Congress Iran–Contra Investigation, testimony of Adolfo Calero, 20 May 1987.
4. US Congress Iran–Contra Investigation, testimony of General Singlaub, 21 May 1987.
5. See also 'North Hired British Mercenary, Singlaub Alleges', *LA Times*, 22 May 1987.
6. *World in Action*, Island at War, 1987.
7. FCO 37/4706, F52, BHC Colombo to FCO London, 24 May 1987.
8. US Congress Iran–Contra Investigation, testimony of Oliver North, 13 July 1987.
9. US Congress Iran–Contra Investigation, testimony of Oliver North, 14 July 1987.
10. *World in Action*, 25 July 1988.
11. FCO 99/2620, 'Nicaragua: Letter from Mr George Foulkes MP to Prime Minister', addendum by P. R. Fearn on 16 March 1987.
12. FCO 99/2620, 'Nicaragua/Blowpipes: Arms for the Contras', P. R. Fearn to Private Secretary on 25 March 1987.

13. Daniel Sanderson, 'Details of Margaret Thatcher's Son Sir Mark's Controversial Business Deals in Oman to Remain Secret Despite National Archives Rule', *Herald*, 21 July 2016.
14. FCO 8/5503, 'Oman/UK: Bilateral Relations', R. J. Dalton to R. A. M. Hendrie, 4 September 1984.
15. FCO 8/5522, 'World in Action Programme on Oman', from D. K. Haskell in London to D. Slater in Muscat, July 3, 1984: 'We have heard from No.10 that relatively little is likely to be said about the cementation contract and Mr Mark Thatcher, but there will apparently be a good deal on Brigadier Landon and his business activities.'
16. FCO 8/5515, 'Oman', letter from Robin Butler to Sir Antony Acland, 24 February 1984.
17. Smith, *The Reluctant Mercenary*, p. 269.
18. Smith, *The Reluctant Mercenary*, p. 279.
19. FCO 37/4706, F62, 'KMS', David Gladstone to FCO London, 3 December 1987.
20. FCO 37/4706, F58, 'KMS', David Gladstone to FCO London, 23 September 1987.
21. FCO 37/4706, F62, KMS, David Gladstone to FCO London, 3 December 1987.
22. Saladin Security website, 'About Us' (accessed 15 February 2015), URL no longer available.
23. Companies House archive, 'Saladin Security Ltd, Directors Report for Year End 30 June 1987'.
24. Companies House archive, 'Saladin Security Ltd, Directors Report for Year End 30 June 1988'.
25. Hansard, 10 March 1987.
26. Companies House, 'Saladin Holdings Ltd and Subsidiaries, Directors' Report for Year Ended 30 June 1997'.
27. United Nations Security Council, S/2015/801. I am grateful for the assistance given by Faduma Hassan in researching Saladin's work in Somalia.
28. Rachel Williamson, 'Securing Oil in Somaliland', 29 September 2014, https://rachelcwilliamson.wordpress.com/2014/09/29/securing-oil-in-somaliland/ (accessed 6 June 2019).
29. Paul Koring, 'Hired Gunmen Protect VIPs', *Globe and Mail*, 22 October 2007.

30. 'EU's €100 Million Afghanistan Security Contract in Limbo, 26 September 2018', *IntelligenceOnline*, www.intelligenceonline.com/corporate-intelligence/2018/09/26/eu-s-e100-million-afghanistan-security-contract-in-limbo,108325302-art, and 'Amarante and Garda Land €85 Million Kabul Prize', *IntelligenceOnline*, www.intelligenceonline.com/corporate-intelligence/2018/11/07/amarante-and-garda-land-e85-million-kabul-prize,108331183-art (accessed 6 June 2019).

31. 'Saladin in Dire Straights in Riyadh', *IntelligenceOnline*, 14 February 2018.

Epilogue

1. 37/3132, 'Sri Lanka: Internal Security', J. P. P. Nason to FCO SAD, 16 December 1983.

2. Letter from David Walker of Saladin Security to STF commandant Deputy Inspector General Mr M. R. Latiff, 30 August 2017.

3. FCO response to a freedom of information request by the author, 11 August 2017, ref: 0579-17.

4. International Truth and Justice Project, 'The Special Task Force', April 2018, www.itjpsl.com/reports/special-task-force (accessed 6 June 2019).

5. The total Police Scotland project cost for training the STF in 2016–17 was £273,718.

6. Phil Miller and Lou MacNamara, 'Undercover Footage Shows British Police are Training Riot-Cops Linked to War Crimes in Sri Lanka', *VICE*, 7 December 2017, www.vice.com/en_uk/article/qv338m/undercover-footage-shows-british-police-are-training-riot-cops-linked-to-war-crimes-in-sri-lanka (accessed 6 June 2019).

7. 'Asian Human Rights Commission Urges Action Over Tamil Boy Killed by Sri Lanka's STF', *Tamil Guardian*, 11 August 2017, www.tamilguardian.com/content/asian-human-rights-commission-urges-action-over-tamil-boy-killed-sri-lankas-stf (accessed 6 June 2019). AHRC spell his name as Sathasivam Madisam – transliterations of Tamil names vary.

8. 'Hundreds of STF Troops and Police Deployed in Thunnalai Crackdown', *Tamil Guardian*, 5 August 2017, www.tamilguardian.com/content/hundreds-stf-troops-and-police-deployed-thunnalai-crackdown (accessed 6 June 2019).

Bibliography

Books

Allen Jr, Calvin H. and Rigsbee II, W. Lynn, *Oman under Qaboos: From Coup to Constitution, 1970–1996* (Frank Cass, London, 2000).

Andrew, Christopher, *The Defence of the Realm, The Authorized History of MI5* (Penguin, London, 2010).

Asher, Michael, *The Regiment: The Real Story of the SAS* (Penguin, London, 2008).

Barber, Noel, *The War of the Running Dogs: How Malaya Defeated the Communist Guerrillas 1948–1960* (Cassell, London, 2004).

Beasant, John, *Oman: The True-Life Drama and Intrigue of an Arab State* (Mainstream Publishing, Edinburgh, 2013).

Cadwallader, Anne, *Lethal Allies: British Collusion in Ireland* (Mercier Press, Cork, 2013).

Cale, George, *My Enemy's Enemy, The Story of the Largest Covert Operation in History: The Arming of the Mujahideen by the CIA* (Atlantic Books, London, 2003).

Cobain, Ian, *Cruel Britannia: A Secret History of Torture* (Portobello Books, London, 2012).

Cobain, Ian, *The History Thieves: Secrets, Lies and the Shaping of a Modern Nation* (Portobello Books, London, 2016).

Cole, Roger and Belfield, Richard, *SAS Operation Storm: SAS Under Siege, Nine Men Against Four Hundred* (Hodder Books, London, 2012).

Coll, Steve, *Ghost Wars* (Penguin, New York, 2004).

Conchiglia, Augusta, *UNITA: Myth and Reality* (London, 1990).

Connor, Ken, *Ghost Force: The Secret History of the SAS* (Cassell, London, 2004).

Cooley, John, *Unholy Wars* (Pluto Press, London, 1999).

Curtis, Mark, *Unpeople: Britain's Secret Human Rights Abuses* (Vintage, 2004).

Devine, Ambassador Frank J., *El Salvador: Embassy Under Attack* (Vantage Press, New York, 1981).

Dickens, Peter, *SAS: Secret War in South East Asia* (Frontline Books, London, 2016).

Dorrill, Stephen, *The Silent Conspiracy: Inside the Intelligence Services in the 1990s* (William Heinemann, London, 1993).

Firmin, Rusty, *The Regiment: 15 Years in the SAS* (Osprey Publishing, Oxford, 2016).

Ghazi, Lieutenant Colonel Mahmood Ahmed, *Afghan War and the Stinger Saga* (Ahmad Publications, Lahore, 2013).

Gladstone, David, *A Sri Lankan Tempest: A Real Life Drama in Five Acts* (self-published, Wotton Underwood, 2017).

Gunaratne, Merril, *Cop in the Crossfire* (Colombo, 2011).

Harrison, Frances, *Still Counting the Dead: Survivors of Sri Lanka's Hidden War* (Portobello Books, London, 2012).

Hart-Davis, Duff, *The War That Never Was: The True Story of the Men Who Fought Britain's Most Secret Battle* (Arrow Books, London, 2012).

Horsfall, Robin, *Fighting Scared* (Cassell, London, 2002).

Kempe, Frederick, *Divorcing the Dictator: America's Bungled Affair with Noriega* (G. P. Putnam's Sons, New York, 1990).

Kerr, Alistair, *Betrayal: The Murder of Robert Nairac GC* (Liverpool Academic Press, Liverpool, 2015).

MacKenzie, Alastair, *Special Force: The Untold Story of 22nd Special Air Service Regiment* (I.B.Tauris, London, 2011).

MacKenzie, Alastair, *Pilgrim Days: From Vietnam to the SAS* (Osprey, Oxford, 2019).

Malathy, N., *A Fleeting Moment in My Country: The Last Years of the LTTE De-Facto State* (Clear Day Books, Atlanta, 2012).

McNab, Andy, *Immediate Action* (Corgi, London, 1996).

Miller, Anthony G. and Morris, Miranda, *Plants of Dhofar, the Southern Region of Oman: Traditional, Economic and Medicinal Uses* (The Office of the Adviser for Conservation of The Environment, Diwan of Royal Court, Muscat, 1988).

Ministry of Information and Culture (Oman), *Oman: A Nation Builds its Future*, (Ministry of Information and Culture, Muscat, 1976).

Ministry of Information and Tourism (Oman), *Oman in Colour* (Ministry of Information and Culture, Muscat, 1974).

Moorcraft, Paul, *Total Destruction of the Tamil Tigers: The Rare Victory of Sri Lanka's Long War* (Pen & Sword, Barnsley, 2012).

Morton, John Percival, *Indian Episode* (unpublished memoir, presented to the British Library in 1982).

Murray, Raymond, *The SAS in Ireland* (Mercier Press, Dublin, 1990).

Pawson, Lara, *In the Name of the People: Angola's Forgotten Massacre* (I.B.Tauris, London, 2014).

Peterson, J. E., *Oman's Insurgencies: The Sultanate's Struggle for Supremacy* (Saqi Books, London, 2007).

Ponnambalam, Satchi, *Sri Lanka: The National Question and the Tamil Liberation Struggle* (Zed Books and Tamil Information Centre, London, 1983).

Ryder, Chris, *The RUC 1922–2000: A Force under Fire* (Arrow Books, London, 2000).

Sivanandan, A., *When Memory Dies* (Arcadia Books, London, 1997).

Sivanandan, A., *Catching History on the Wing* (Pluto Press, London, 2008).

Sivanayagam, S., *Sri Lanka: 10 Years of Jayewardene Rule* (Tamil Information and Research Unit, Madras, 1987).

Smith, Tim, *The Reluctant Mercenary: The Recollections of a British Ex-Army Helicopter Pilot in the Anti Terrorist War in Sri Lanka* (Book Guild, Lewes, 2002).

Stalker, John, *The Stalker Affair: The Shocking True Story of Six Deaths and a Notorious Cover-up* (Viking, London, 1988).

Takriti, Abdel Razzaq, *Monsoon Revolution: Republicans, Sultans and Empires in Oman, 1965–1976* (Oxford University Press, Oxford, 2016).

Urban, Mark, *Big Boys' Rules: The SAS and the Secret Struggle against the IRA* (Faber and Faber, London, 1992).

Urban, Mark, *War in Afghanistan* (St Martin's Press, New York, 1990).

Urwin, Margaret, *A State in Denial: British Collaboration with Loyalist Paramilitaries* (Mercier Press, Cork, 2016).

Whitaker, Mark P., *Learning Politics from Sivaram* (Pluto Press, London, 2006).

Reports and Journals

Halliday, Fred, 'The Ceylonese Insurrection', *New Left Review*, vol. 1, no. 69, September/October 1971.

International Truth and Justice Project, 'The Special Task Force', April 2018.

NorthEast Secretariat On Human Rights, 'Piramanthanaru Massacre', April 2006.

NorthEast Secretariat On Human Rights, 'Lest We Forget, Massacres of Tamils 1956–2001, part 1', 2007.

People's Liberation Organisation of Tamil Eelam, *Spark*, vol. 1, no. 2, January 1985, Madras, India.

President's Special Review Board (Tower Commission), 26 February 1987.

Index

Index

Index

Morton, John Percival, 71–4, 76, 95, 100, 102, 108
Mossad, 1
MPLA, *see Popular Movement for the Liberation of Angola*
Mujahideen, 177–80, 182, 189
Mullaittivu, 210
Musandam, 84
Muscat, 12–15, 17–18, 77, 84, 90, 255, 257, 289, 291
Muslim, 15–16, 64, 148, 158–62, 165, 276–7, 283–5
Muttur, 278

Nairac, Robert, 88–9
Nairn, Donald, 84
Nairobi, 164, 270–1
Namibia, 104
Napalm, 54–5, 248, 275
Napier, Shirley, 235–6
Nason, Justin, 147, 150
Nasser, Gamal Abdel, 6–8, 13, 25, 79
National Archives UK, 9, 55, 72, 134, 140, 149, 176, 179, 234, 256
National Army Museum, 9, 80–2
National Endowment for Democracy, 287
National Intelligence Bureau, 166
National Rifle Association, 287
Necklacing, 154, 186
Nedunkerni, 186
Netherlands, 9, 44–5, 47
Nepal, 26, 272
Newham, 239
New Zealand, 84, 258–9
Nicaragua Solidarity Campaign, 254
Nicaraguan Democratic Coordinator, 114
Nicaraguan Democratic Force, 112–13, 115–21, 142–6, 215, 217–19, 237, 243, 246–7, 249–53
Nightingale, Andrew, 78, 265
and death in car crash, 86–9
Non-Aligned Movement, 202
North, Oliver, 116–21, 143–4, 177, 180, 215, 217–19, 236–7, 243, 246–52, 287

Northern Ireland, 104, 164, 184, 232, 272, 290
and Bloody Sunday, 23, 170
and Flagstaff hill incident, 33–5, 38–9, 124–5
and sharing experience with Sri Lanka, 95, 97, 100, 130–1, 199, 277, 279
Northern Ireland Office, 35, 97, 146

Oman, 9, 12–19, 30, 32, 58, 77–8, 81, 83–5, 90–1, 101, 104–8, 256–9, 267, 285, 287–91
and Internal Security Service, 289–90
and National Security Council, 290–1
and Northern Frontier Regiment, 290
and Palace Office, 258–9
and Royal Guard Brigade, 17
and Royal Office, 291
and Special Task Force, 290
and transfer of SAS from Oman to Northern Ireland, 33–5
Omanization, 16–17
Operation Contravene, 35
Operation Liberation, 248
Operation Southern Comfort, 85
Operation Storm, 33, 39–42
Operation Thalib, 258
Ortega, Daniel, 114, 286–7
Osamor, Kate, 284
Ovenden, Ken, 21
Owen, Robert, 143–4
Oxford University, 82–3, 99, 101

Pakistan, 98, 179
Pakiyanathan, Ponnuthurai, 1
Palaly airbase, 196, 210, 241–2, 244–5
Palanivel, Sabaratnam, 96
Palestine, 102, 104
solidarity from Dhufar, 85
solidarity from Tamils, 159
Palk Strait, 99
Pan-Arabism, 6, 159
Panjshir Valley, 181

Index

331

Index

Trafalgar House, 256
Trincomalee, 59, 96, 195–6, 199–200, 202, 204, 221, 227, 231, 233–5, 265, 278, 298–9
TULF, *see Tamil United Liberation Front*
Tulliallan Castle, 282

Uganda, 9, 45–6, 91, 106
United Arab Emirates, 31
Uruguay, 9, 92–3, 106
Uzbekistan, 182

Vaddukoddai resolution, 64, 68
Valvettithurai, 67, 229–30
Varadakumar, Varaimuttu, 190, 238–9
Vavuniya, 96, 186, 203, 221, 232
Victoria, Queen, 7
Vietnam, 101, 176, 191, 214, 258
VVT, *see Valvettithurai*

Waddington, David, 239–40
Wakefield, Peter, 20–1
Walker, David, 180
  and Argentina, 42–4
  and army reservist, 51
  and councillor, 91–3, 115–16, 122, 129, 145, 168, 197, 207, 254
  and KMS founder, 78
  and Oman, 39–42
  and National Army Museum, 81
  and Nicaragua, 115–18, 120, 144–6, 217–18, 236–7, 246–52
  and questions in Parliament, 254–7
  and Saladin, 265–71, 279–80

and Sri Lanka, 138, 155, 162, 191–3, 205, 220, 225, 240, 243–6, 279–80
Walton-on-Thames, 92, 129
Ward Place, 207
Warsaw, 43
Waterloo, 169
Weeratunga, Tissa, 73, 166
Welsh Guards, 79, 173
Westland Helicopters, 139
Weybridge, 92, 145
Wharton, Dave, 210, 220
White, Dave, 291
White House, 115, 236, 246
White, Ken, *see Baty, Brian*
White Sultan, *see Landon, Tim*
Whyte, Ken, *see Baty, Brian*
Wijesundara, Palinda, 281, 285
Wijesuria, Zerni, 167
Willpattu, 242
Wilson, Harold, 23–7, 35, 50, 88
Wilson, Willie, 264
Wine glasses, 232, 234
Wingate Gray, Brigadier Mike, 19, 78, 155, 265, 271
Wonga Coup, 273

Yamani, Sheikh Ahmed Zaki, 31
Yellow card, 35
Yemen, 5, 7, 12–13, 19, 25, 37, 77–80, 179, 199, 271
Young, Janet, 200, 241

Zambia, 31
Zeek, 78
Zion Church, 162